20th Topical Conference on Gas Utilization 2020

Topical Conference at the 2020 AIChE Spring Meeting and 16th Global Congress on Process Safety

Online
17 – 21 August 2020

ISBN: 978-1-7138-2322-3

Printed from e-media with permission by:

Curran Associates, Inc.
57 Morehouse Lane
Red Hook, NY 12571

Some format issues inherent in the e-media version may also appear in this print version.

Printed with permission by Curran Associates, Inc. (2021)

For permission requests, please contact AIChE
at the address below.

AIChE
120 Wall Street, FL 23
New York, NY 10005-4020

Phone: (800) 242-4363
Fax: (203) 775-5177

www.aiche.org

Additional copies of this publication are available from:

Curran Associates, Inc.
57 Morehouse Lane
Red Hook, NY 12571 USA
Phone: 845-758-0400
Fax: 845-758-2633
Email: curran@proceedings.com
Web: www.proceedings.com

TABLE OF CONTENTS

591161 2020 AIChE Spring Meeting & 16th GCPS AGILE Award Keynote Address - Beyond the Horizon: The Factors and Forces Affecting the Global Petrochemical Industry

Abstract:

Over the past 10 years, LyondellBasell and the global petrochemical industry has undergone a dramatic transformation. In this era of unprecedented disruption, it is likely that the next the decade will present both challenges and opportunities for the petrochemical sector. LyondellBasell CEO Bob Patel will summarize the company's progress to date and provide a forward-looking perspective on key opportunities and risks faced by the industry in 2020 and beyond.

591157 Securing Enterprises via Cybersecurity Understanding, Resiliency, and Education

Abstract:

Securing Enterprises via Cybersecurity Understanding, Resiliency, and Education

Industry 4.0 initiatives are enabled by leveraging the deployment of information technologies into the operational technology environment. While a logical approach in defining the future of smart manufacturing, it remains fret with challenges and risks that must be mitigated assure enterprise resilience to cyber and cyber-physical attacks. As economies recalibrate in the aftermath of the current public health emergency, definitions of what constitutes "critical infrastructure" are broadening while, at the same time, attack surfaces and the threat landscape are expanding. This means companies must place a greater emphasis on taking the steps and making the investments necessary to overcome the challenges and navigate these challenges and risks to maximize their resilience and secure and protect their enterprises. This presentation will explore the link between cybersecurity and process safety in the context of a safety culture. It will also identify cybersecurity challenges in industry 4.0 initiatives, the integration of information technologies (IT) and operational technologies (IT), the delivery of training, and the need for credentialing and ongoing professional development.

The Techno-Economic Prospects of Using Piperazine for Carbon Dioxide Removal from Natural Gas

Mohammed I. Bargo*, Mustafa M. Ali, Nayer M. Hassan, Hussein S. Ayoub
Department of Chemical Engineering, Faculty of Engineering, University of Khartoum
Khartoum, Sudan

Abstract

One technology used in the removal of carbon dioxide from natural gas is the absorption process, in which Alkanolamines are habitually used as solvents. Since the 1960's and 70's several Amines have come into general use, however, limited information has been reported in the literature concerning the Amine best suited to treat sweet natural gas. From the absorption rate point of view, primary or secondary Amines (e.g. MEA) are preferred. However, in case the costs of regeneration are also taken into account, tertiary Amines (e.g. DIPA) are much more attractive.

The core objective of this work is to evaluate the potential of using Piperazine against the most common Amines that are used nowadays. This article presents a novel investigations of the most significant effective parameters for processing Piperazine, and their influences on the plant efficiency and investment cost. And also has demonstrated that, choosing Piperazine (30%wt) could reduce more than 10% of the operating cost and 4% of the equipment cost in the Acid Gas Removal Unit against the other solvents.

Results show that Piperazine (30% wt.) is the most effective solvent, which performs with effectiveness of 0.8 mol-CO2absorbed/mol-Amine at the specified operating conditions. Moreover it has preceded other alternatives in both operating and equipment cost points of view. Results also show that, the concentration of the Amines has significant effect on the operating cost of this unit: some of them exhibit increasing trends, others decreasing trends, and others even have inflection point (concave up curve).

Keywords: Natural gas; CO2 removal; Piperazine; Aspen HYSYS

Introduction

Natural gas will be an enabler for key game-changing renewable technologies not just solar, wind and so forth as a complementary source but also in working through making hydrogen a reasonable source. Natural gas is a necessary bridge fuel that will be a critical part of shifting the energy mix away from coal.

Carbon dioxide is an impurity present in all natural gas reservoirs all across the world in various amounts. This CO2 is an inert component, so it reduces the energy content of the natural gas and hence it's selling price. Moreover it becomes acidic and corrosive at the presence of water, which has the potential to damage the pipeline and the equipment systems. As the global demand of natural gas is expected to grow at a rate of 1.6% a year until 2024 [1], there will be more carbon dioxide waiting to be removed.

Gas absorption process using Alkanoamine to remove CO2 have been in use since the 1930. In earlier applications, these amines were represented by MEA and DEA. Later, other amines like DGA and DIPA gained commercial recognition [2]. In recent years, various MDEA and specialty solvents blends were developed (mostly using Piperazine and MEA) which take advantage of the low-energy characteristics of MDEA.

From the absorption rate point of view, primary or secondary amines (e.g. MEA) are preferred. However, in case the cost of regeneration are also taken into account, tertiary amines (e.g. DIPA) are much more attractive. As more effective solvent is used, the less amount of it will be needed, but in contrast there will be a high cost to regenerate it. Moreover, 50 to 70% of the initial investment for an

* Corresponding author:
E-mail addresses: mohammedbargo@gmail.com

acid gas removal unit is directly associated with the magnitude of the solvent circulation rate, and 70% of the operating cost result from the regeneration of this amount [3]. Therefore, there is a significance of using the proper solvent.

Piperazine (PZ) is a di-amine has been used in several DOE (Department of energy, USA)-funded capture projects as a $CO2$ capture solvent because it has a higher cyclic capacity, faster kinetics, is thermally stable at higher temperatures and requires less energy to strip $CO2$, resulting in lower regeneration energy. Bishnoi et al conducted modelling experiments and revealed that the carbon dioxide removal rate is more for 8m PZ amine compared to the other amines like MDEA, MEA, DEA and their blends [4].

However, PZ forms a solid phase with water (PZ.6H2O) and also with CO2 (H+PZCOO-H2O) which can cause precipitation issues and problems for plant operators, but Rochelle et al have developed a solubility envelope show the permitted concentrations with a specified operating conditions and examined AMP as promising additive [5].

The core objective of this work is to evaluate the potential of using PZ against the most common Amines which are used nowadays.

Theory

The reactive absorption of CO2 at the gas-liquid interface involves the diffusion of CO2 from the bulk gas phase to the interface, reaction at the interface and the diffusion of aqueous CO2 species away from the interface into the bulk liquid.

The flux of CO2 is given as:

$$N = K_G (P_{co2} - P^*_{co2})$$

Where N is mass-transfer rate, K_G is the overall mass-transfer coefficient, Pco2 and P*co2 represent CO2 partial pressures in the bulk gas and liquid respectively.

The overall mass-transfer coefficient is depicted graphically in Figure I and can be expressed in terms of the gas-film mass transfer coefficient k_g and liquid-film mass transfer coefficient k_{lo} (physical diffusion) as:

$$\frac{1}{K_G} = \frac{1}{k_g} + \frac{H_{co2}}{E * k_{lo}}$$

Where E represents the enhancement of the liquid-film mass transfer rate due to the reaction, H_{co2} is the Henry's law constant, and the superscripts refer to similar quantities as above.

Solvent which can react faster with CO2 are benefited for CO2 capture because the mass transfer coefficient is higher. Similarly, solvents where CO2 can diffuse more readily (i.e. less viscous solvent) also lead to higher E factor and higher mass transfer coefficient.

CO2 can react with amines in two different ways: carbonate formation (react with primary and secondary) and bicarbonate formation [6].

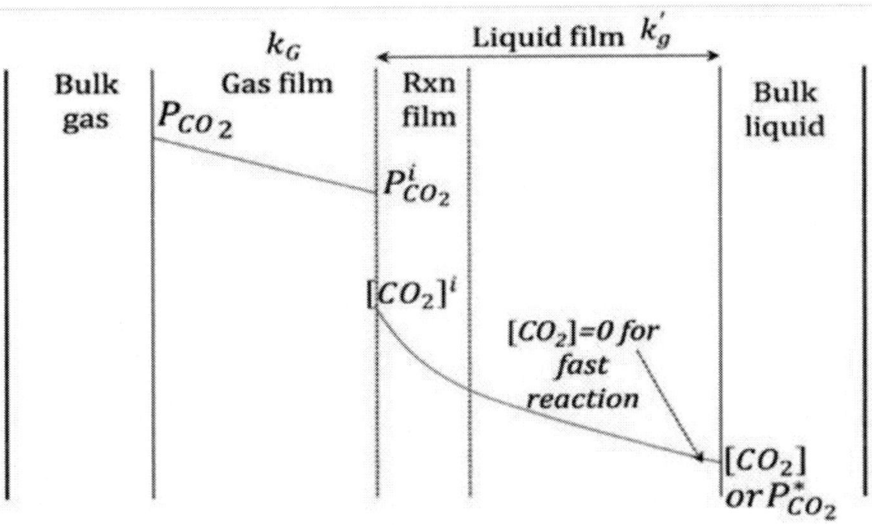

Figure I Schematic of CO2 Absorption into a Liquid Solvent with Chemical Reaction [8]

Methodology

The specifications used for this study were taken from the results in an ongoing project to recover associated natural gas which comes off Neem Field in Western Sudan. Gas flow rate of 22 MMSCFD comes in pressure of 50 bara, and temperature of 40 °C.

Table (1) shows the specification which is considerably sweet for the gas, also shows the targeted product to meet the pipeline requirements.

Table 1 Feed gas composition [7]

Component	Mol%
Nitrogen	0.6
CO_2	1
H_2S	0
H2O	0.9
Methane	72
Ethane	13
Propane	6
i-Butane	0.9
n-Butane	2
i-Pentane	0.7
n-Pentane	0.8
C6+	1.7

The target is to achieve the same quality of 0.05%mol of carbon dioxide.

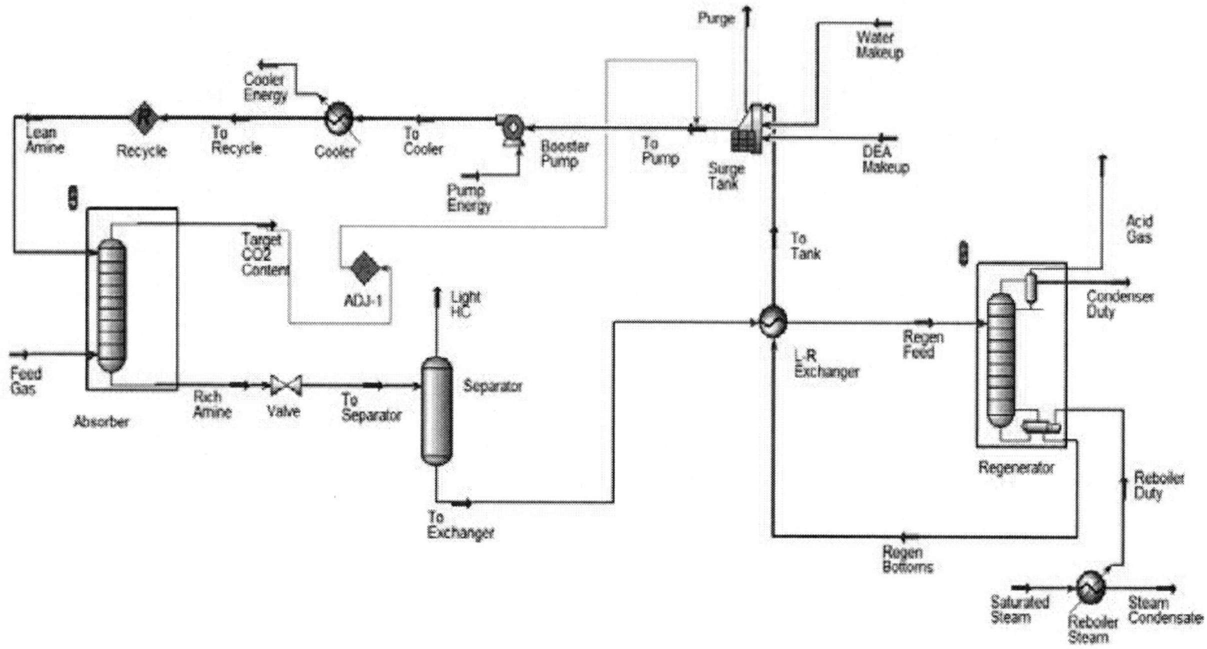

Figure II Acid gas removal unit simulation

Regarding the methodology, firstly six solvents had been identified to be investigated, then an acceptable range of concentrations for each solvent is defined. Next, a design of the facilities at each concentration for the six solvents are simulated using Aspen HYSYS software, see Figure II.

After that, the most economic concentration of each solvent is selected, and then the chosen concentrations are compared in performance and economic point of view.

Specifying acceptable range of concentrations for each Amine requires finding which range may attain solvent stability and reduce foaming tendency and corrosion rates of the solvent. Aspen Process Economic Analyzer (APEA) was used to evaluate the operational and capital cost of this process and material costs.

Results and Discussions

Effect of the Amine concentration on the operating cost:

On pursuing the most economic concentration of each solvent, the effect of the Amine's concentration on the operating cost has been studied, but the Amines have showed three different ways of behavior:

1- Piperazine and DEA : its operating cost increases as the concentration increases (Figure III)

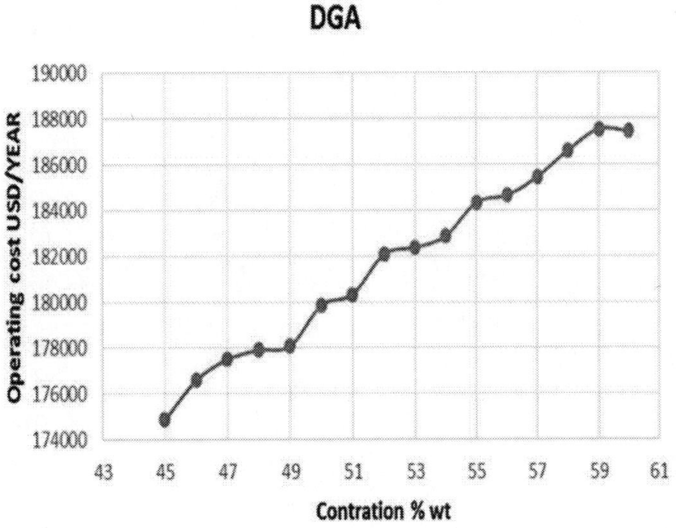

(a)

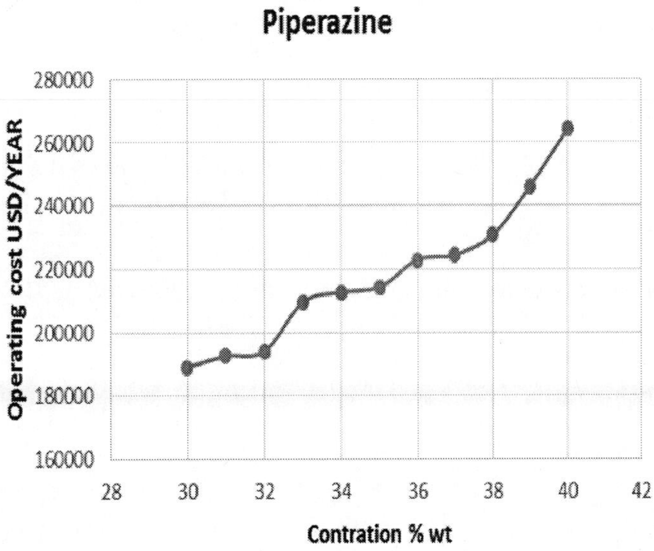

(b)

Figure III Effect of the Amine concentration on the operating cost_the increasing behaviour: a) DGA, b) Piperazine

2- DGA, DIPA and MDEA: their operating cost decreases as the concentration increases (Figure IV)

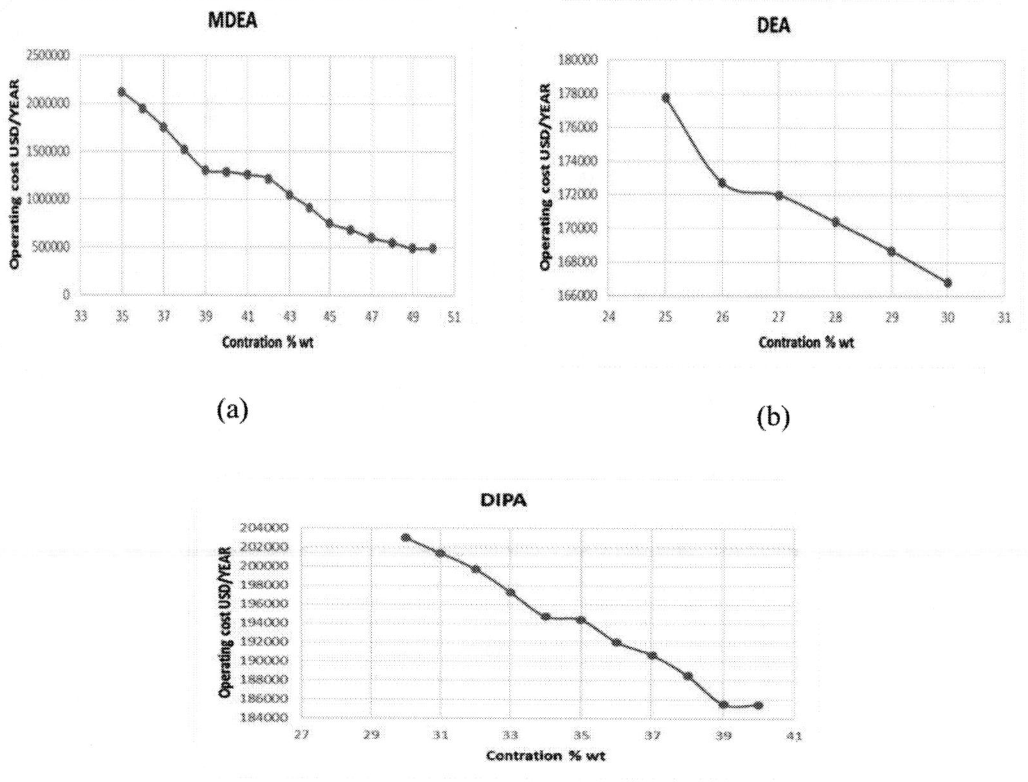

(c)

Figure IV Effect of the Amine concentration on the operating cost_the decreasing behaviour: (a) MDEA, (b) DEA and (c) DIPA.

3- MEA: has inflection point in its curve that is concaved up (Figure V).

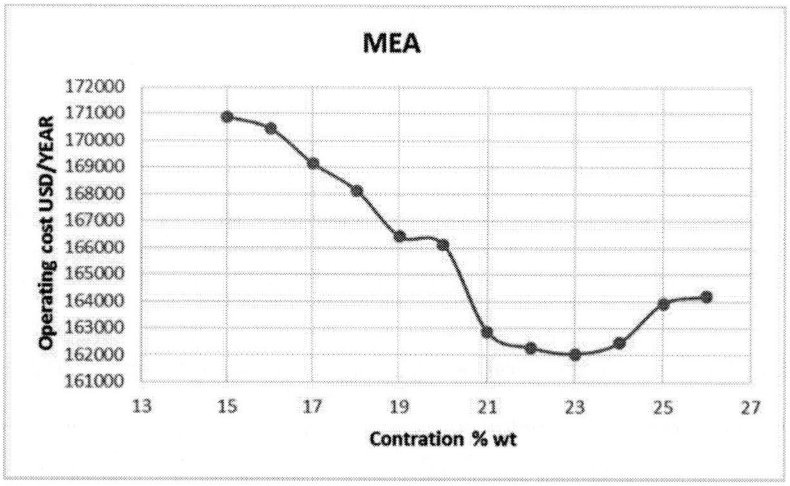

Figure V Effect of MEA concentration on the operating cost of acid gas removal unit.

This behavior is due to the fluctuation of the operating cost which concerns the reboiler duty cost and the makeup cost. For each solvent, there is a certain limit to which distillation column can recover the Amine concentration; because reboiler temperature cannot exceed the degradation temperature of the solvent; hence the makeup must carry the shortening of the distillation to achieve the targeted Amine-water solution concentration.

Performance as basis of comparison:
How many moles of CO2 could be absorbed by one mole of Piperazine compared to the other solvents is shown in Figure VI:

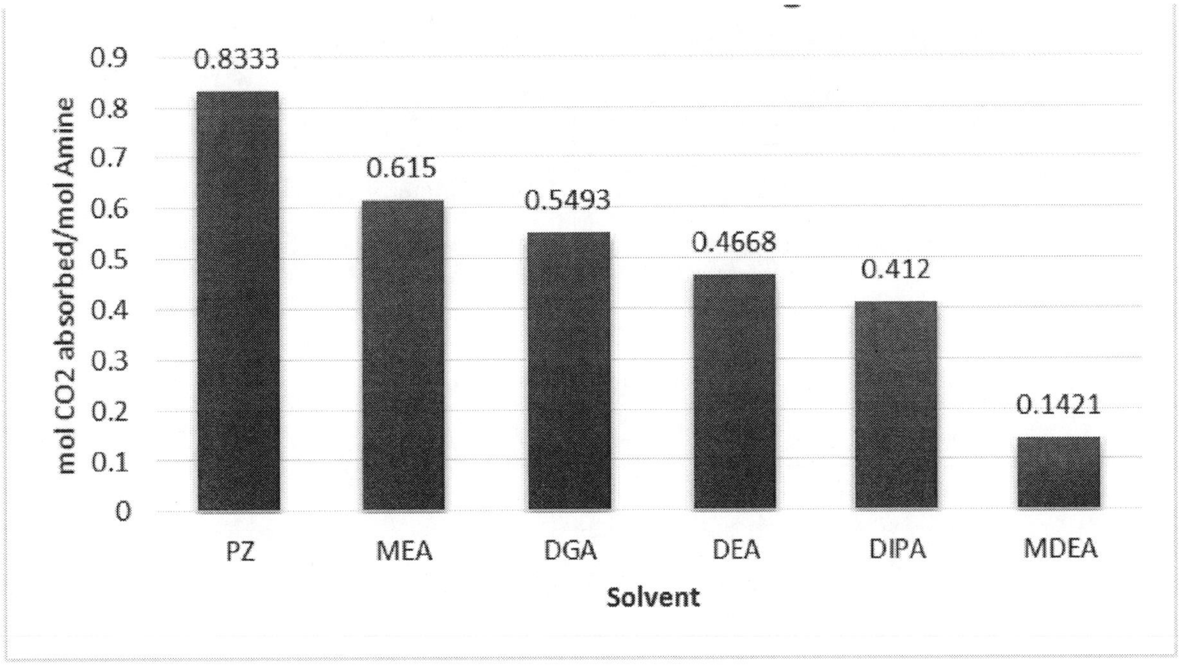

Figure VI The effectiveness for each solvent

Economic basis of comparison:
It's clear that Piperazine preceded other alternatives in both: the operating and equipment costs as shown in Figure VII & Figure VIII:

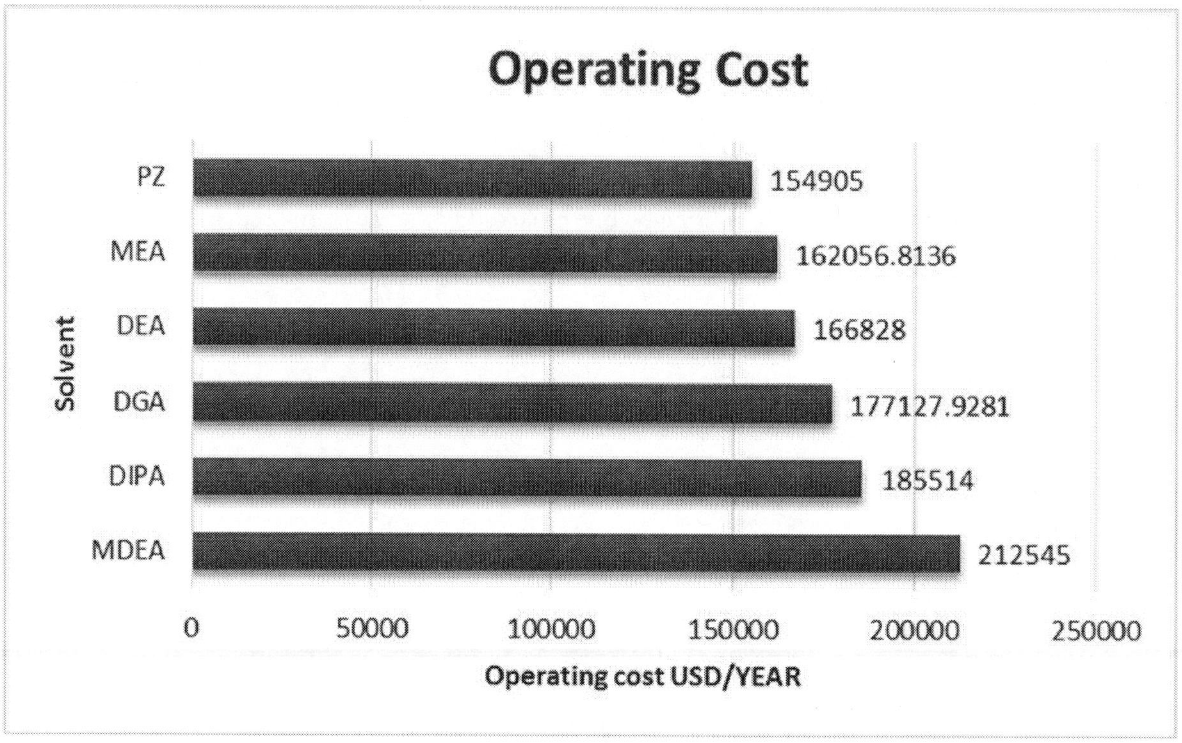

Figure VIII Operating cost of the acid gas removal unit for each solvent

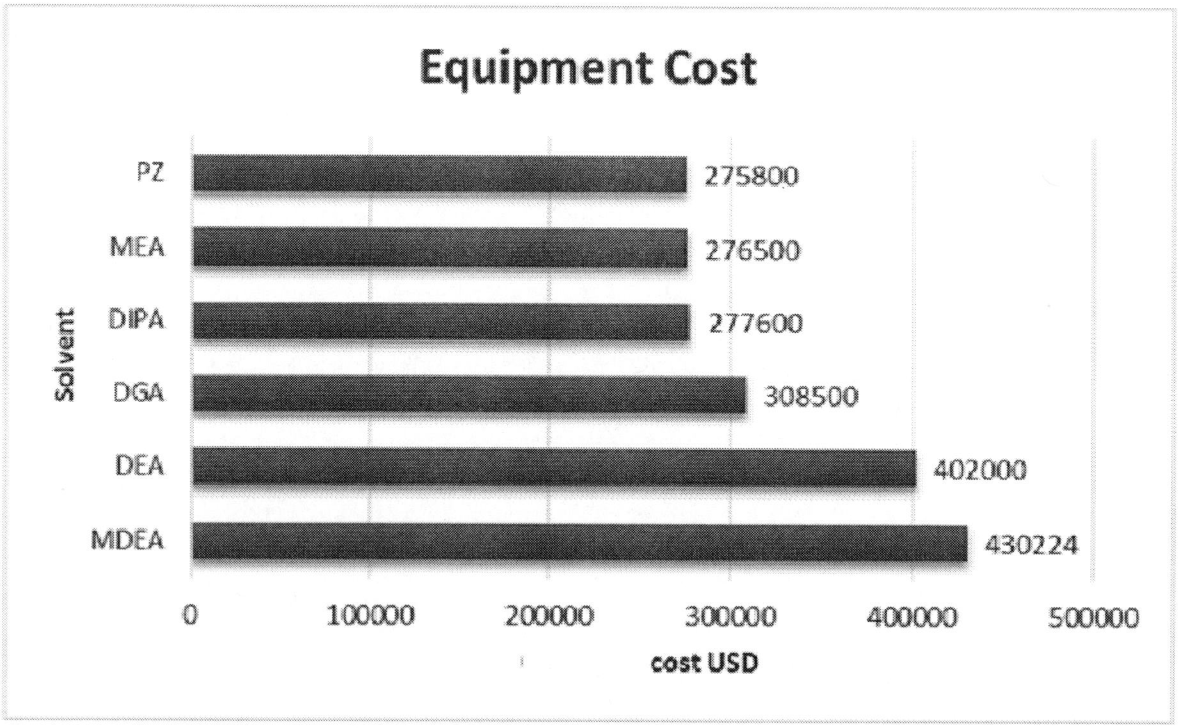

Figure VII Equipment cost of the acid gas removal unit for each solvent

The effect of the Piperazine's concentration on the reboiler duty and the makeup flow rate: Figure IX & Figure X show variations in the energy consumption of the reboiler and makeup addition flow rate to achieve the required concentration.

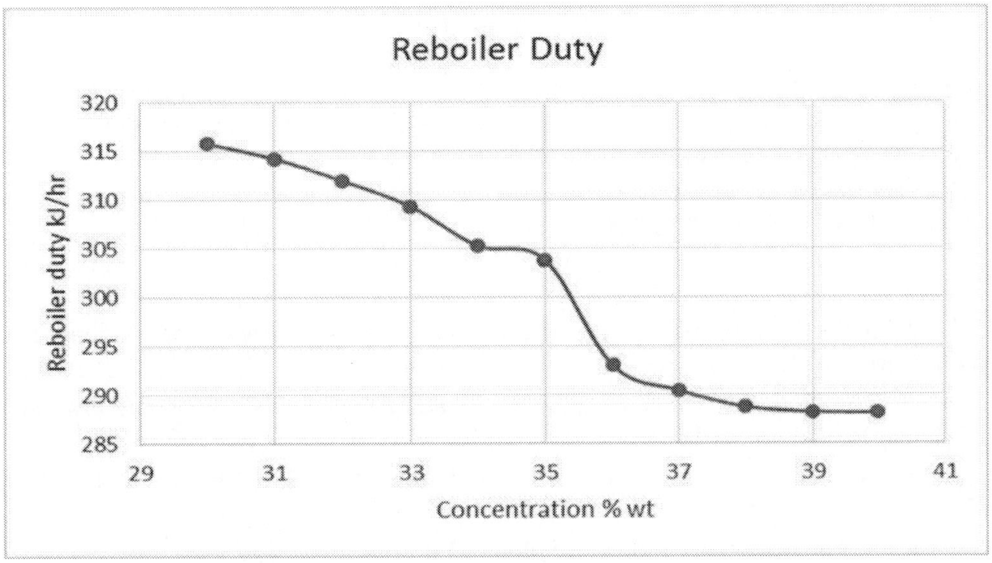

Figure IX Effect of Piperazine concentration on the rebuilder duty

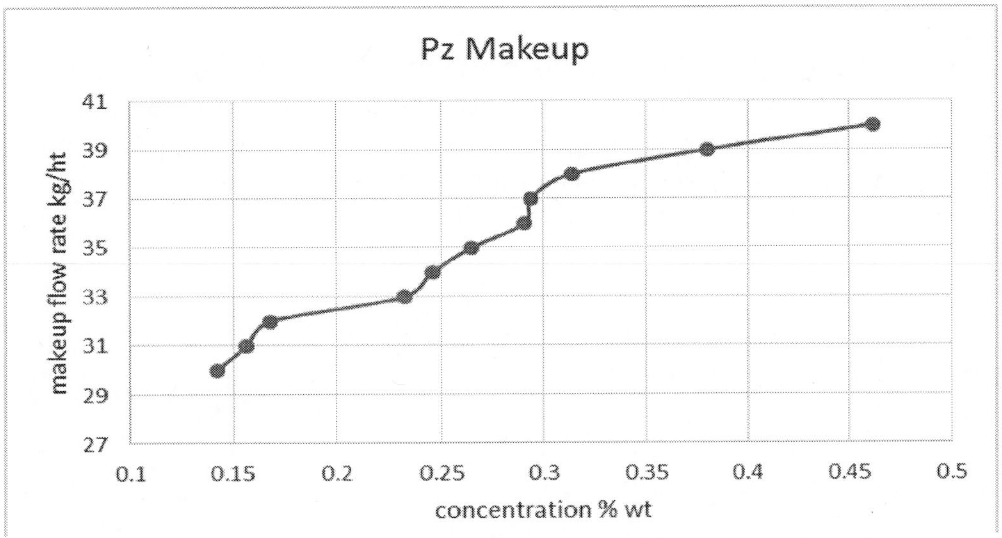

Figure X Effect of Piperazine concentration on the Piperazine makeup flow rate

Conclusion

- Piperazine 8m (30% wt.) is efficiently and effectively the best solvent for this operation, and whether choosing it or not may cost +10% in the operating cost and +4% in the equipment's cost compared to MEA (23% wt.).
- Piperazine 8m (30% wt.) preformed in 0.8 mol.CO_2 absorbed/ mol Amine at 50 bara and temperature of 40 °C which is the strongest recorded result.
- The fluctuation of the operating cost with the different solvents result from the restrictions of the degradation temperature of each solvent which limits the capability of the stripping column.
- Choosing the optimum concentration for Piperazine may reduce 23% of the improper concentration operating cost.
- The market price of the Piperazine may determine its best concentration, since the makeup is directly proportional to it.

Acknowledgements

We would like to thank our MSc. Siddig Sied Ahmed who offered substantial support and for his warmth encouragement and guidance. In addition, we would like to thank (Sudapet Oil& Gas Co. Ltd.) to sponsor and offer technical support for this project. Special recognition to MSc. Motaz Siddig (U of K Oil &Natural Gas Engineering Department) for his guidance during the initial stages of the project, Eng. Ayman Musa for their help.

References

[1] EIA, "Gas 2019 - Executive summary, Analysis and forecast to 2024," International Energy Agency, Paris, 2019.

[2] S. Mokhatab, W. Poe and J. Mak, Handbook of Natural Gas Transmission and Processing, Gulf Professional Publishing, 2015.

[3] Astarita, D. W. S. G. and A. Bisio, Gas Treating with Chemical Solvents, J. Wiley and Sons, 1983.

[4] B. S., "Carbon Dioxide Absorption and Solution Equilibrium in Piperazine Activated," The University of Texas at Austin. Ph.D. Dissertation, 2000.

[5] H. Lib, L. Li, T. Nguyen, G. T. Rochelle and J. Chen, "Characterization of Piperazine/2-Aminomethylpropanol for Carbon Dioxide Capture," *Energy Procedia,* vol. 37, pp. 340-352, 2013.

[6] "SOLVENT-BASED CO2 CAPTURE," in *DOE/NETL Carbon Capture Program—Carbon Dioxide Capture Handbook*, National Energy Technology Laboratory, 2015, pp. 14-17.

[7] E. Ismail, "NEEM & DIFFRA Gas Composition," GNPOC , Khartoum, 2015.

[8] R. E. Tsai, "Mass transfer area of structured packing, Masters Thesis," http://repositories.lib.utexas.edu/handle/2152/ETD-UT-2010-05-1412, 2010.

Accelerating Carbon Neutrality with Sustainable Biomass Cycle

Brian P. Kolodji, PE; Black Swan, LLC; Bakersfield, CA

Preface:

Beginning with a quote about the beginning that is the inspiration for this work "...then the Lord God formed the man out of the dust of the ground and blew into his nostrils the breath of life, and the man became a living being..." **(Reference 1**), I want to thank God for the people sent to advance this display of the power of breath as a solution for climate change, specifically Paramjit Dosanjh, Dr. Bruce Kimball, Dr. Brian Marsh, as well as CDFA's Secretary Karen Ross, Chief Scientist Dr. Amit Gunasakara, and Project Manager Scott Weeks whose technical support, funding, and encouragement are indispensable.

Introduction:

Crop Carbon Dioxide (CO2) Enrichment or CCE, discovered over 200 years ago (reference 2), is the agricultural practice of maintaining a CO2 concentration above ambient concentration in a crop's biosphere. The practice of CCE is proven for almost a century (see Darymple, D.G. of USDA from 1973 in reference 2) to benefit crop production. In references 3 and 4, benefits are shown by Dr. Bruce Kimball of USDA to be an average 33% but as high as 200% increase in crop yield while crop water utilization efficiency (WUE), an indication of crop irrigation water saved, also increases an average 48% (where Average WUE Increase=Increased Average Yield/Decreased Average Evapotranspiration =1.33/0.90). In greenhouses, CCE operations have been performed worldwide. In Tehachapi, California there are two world class industrial-scale 32-acre SunSelect (see reference 5) CCE greenhouses used to enhance growth of cucumbers, tomatoes, and chiles crops as provided by flue gas from fit for purpose built natural gas fired boilers. Free Air (meaning without enclosures such as greenhouse or open top chamber) CO2 Enrichment (also known as FACE technology) has been researched for decades by USDA (see reference 4) and Brookhaven National Laboratory. CCE using FACE has been practiced worldwide for crops such as wheat, cotton, grapes, and soybean. The CO2, ultimately diluted down to 550ppm, in conventional FACE technology is provided pressurized (250 psig+) at high purity (90%+ or 900,000+ ppm CO2) costing over $200/ton for CCE. FACE technology has only entailed very small fractions (under 1/10) of a single acre (or 4000 sq feet or 400 sq m) without being profitable, that is, until now.

Black Swan, LLC with FGXB™, FACE technology is the only commercially viable FACE technology for full scale orchards due to inherently lower cost and risk by conditioning/processing CO2 freely emitted at low pressure (well under 1 psig) and at lower/safer CO2 concentrations (5 to 10% or 50,000 to 100,000 ppm) before being reduced down to the required 550 ppm for CCE. This CO2 is extracted from existing retrofitted flue gas stacks in power plants and refineries, the largest source of Greenhouse Gas (60 Billion+ Tons/year or TPY), and has CO2 concentrations similar to respired breath (5% CO2), the natural concentration respired by fauna used for photosynthesis of flora. FGXB™ pilot plants have been operating as early as 2016 (see reference 6) at several orchards (almonds/cherry) throughout Kern County California. This FGXB pilot at the Dosanjh orchard is the largest scale FACE operation to date worldwide at nearly 0.25 acres (10,000 sq feet) utilizing up to 200 tons CO2/year from flue gas and is funded by a California Department of Food and Agriculture (CDFA) State Water Environmental Efficiency Program (SWEEP) Grant awarded in 2018 to Dosanjh Bros (Paramjit Dosanjh.) This facility at the Dosanjh Bros almond orchard is planned to be expanded to a 20-acre, 2000 tree demonstration facility utilizing up to 10,000 Ton CO2/Year, supplied by the neighboring refinery, currently freely emitting from industrial stacks almost 200,000 tons per year of CO2. The planned 2021 demonstration facility will be over 100 times larger than the patented/ patent pending FGXB™ technology piloted as discussed herein.

CO2 in FGXB™ pilots are provided using exhaust from natural gas (methane) and LPG (propane) fuel fired internal combustion engine exhaust (including a 350 HP IC engine driven well pump.) Cooling of the flue gas by radiative heat loss with well piping coils, or sublimation of dry ice, or quench water column have occurred at these pilots. Conditioned flue gas was delivered to trees by vacuum blowers, in-line fans, or a large tractor driven almond blower and delivered to crops in above grade ducting or irrigations systems. The CO2 concentration produced are in excess of 6% (60,000 ppm) CO2 as expected from a refinery natural gas fired 5 MW cogen unit stack gas. This pilot that is the subject of this paper fed 20 trees directly, and up to 50 surrounding trees indirectly, started in Fall 2018 and discontinued in Fall of 2019 after harvest, and then resumed again in June 2020 to current (September 2020) at a smaller scale (directly to 9 trees.) In 2018 to July 2020 flue gas exhaust cooling was provided with dry ice.

The Dry Ice flue gas cooling system was replaced after July 13, 2020 with a water quench column (Figure 1) installed, with operations resuming on August 5, 2020. The patent pending quench column provides cooling for hot flue gas exhaust from exhaust temperature of over 300F to under 100F. The quench column provides much more economic and much higher capacity as needed with industry supplied CO2 from stack gases. Operating costs are much lower in not requiring dry ice, and instead involve running a recirculation pump powered by the portable generator. The recirculated quench water ultimately cools and extracts water vapor and heat. The process makes, and does not consume water and does not accumulate sensible heat from the cooling water due to latent heat absorbed by condensation. In combustion, for every molecule of CO2 produced, at least 2 molecules of water vapor are formed and condensed in this process. Another patented advancement involved replacement of blower/vacuums with quieter, more economic, and leak free in-line duct fans. The flue gas is then sent to tree vents (Figure 2) showing 8" CO2 supply header, branch and 12" diameter supply stack to a tree. To supplement CO2, additional CO2 volumetric flow is provided by gasoline driven internal combustion engines, providing in excess of a rate of 200 tons of CO2 per year.

A process flow diagram/ plot and tree layout (Figure 3) is provided, and can be best understood after reviewing the key in the bottom right corner.

Summary/ Description of Operations:

The subject 2020 Dosanjh Orchard pilot operations involved pre-commissioning; commissioning; operations during maintenance, repair, reconstruction, and startup; and ultimately near normal operations. As expected, every day of operation of the pilot plant involved some significant change/process parameter evaluation/discovery at this point of development, thus the reason it is called a pilot and not yet demonstration facility. These changes involved changing up and/or down of fan capacities (from 300 to 2000 cfm), tree vent size (from 8" to 20"), vent height (from 3 to 9 feet), CO2 supply volumetric rate, CO2 concentration/dilution at the stacks (0.5 to over 15% CO2), pneumatic balancing of volume to each tree, and number of trees applied. Regarding concentration/dilution of CO2 at the stack, this is more of a balancing act between the flue gas supply rate, and the fan size(s), as well as chasing leaks in the system.

Two Independent Operations Data Sets were generated (see references 7 and 8) over the period of July 31 through August 24, 2020. Operations during this period lasted from under an hour to the longest being just over 6 hours. The average run length is between 2 and 3 hours in a day. Highlights of successful CO2 application show concentrations at the 5 foot level above grade at the vent/ CO2 supply outlet between 0.5 to 10% or 5000 to 100000 ppm CO2, with concentration sustained levels at over 5% or 50,000 ppm for over 30 minutes. Concentrations at 10 feet above grade inside the tree canopy were at sustained levels at over 800 ppm for 30 minutes (8/22/2020), peaked/spiked for several minutes at over 1100 ppm on this day, and on 8/21 and 22/ 2020, as well as achieved a maximum spike of 1700 ppm on 8/24/2020 at about 8:30AM. At 15 feet above grade between the trees, a sustained level of 500 to 800 ppm was achieved for 15 minutes on 8/21/2020 (see Figure 4), with spikes between 500 and peaking at 1700 ppm for several minutes on 8/24/2020. Concentrations at the 10 foot and 15 foot level were inverted on several occasions, where the concentration at 15 foot level was higher than at the 10 foot level, and is usually an indicator of overdose of trees from excessive CO2 concentration and/or rates. This phenomenon occurred commonly, at least once, if not several times a day, as identified on Figures 7 thru 17.

To assess each process parameter (temperature, flow rate, CO2 concentration, air leakage, CO2 dispersion, soil moisture, tree water utilization efficiency, crop yield) analytical and process instruments are placed in selected process locations for measurement and analysis of process performance. The following are a detailed description of operational modes, parameters, and measurements based on the Process Flow Diagram in Figure 3. Two primary operational modes exist for the pilot operations based on cooling. The operational mode performed in June/July 2020 used Dry Ice only (and not the quench column) for Cooling (with Valves 3A/3B/4C Closed/Open/Closed.) The operational mode described in the balance of this document describes operations in August/ September 2020, and uses the quench column only (with valves 3A/3B Open/Closed.)

CO2 is a component of the flue gas generated by flue gas generators usually at a concentration above 5% (50,000 ppm) and is provided via primary and optional flue gas generators. The primary generator is the portable LPG (propane) fired electrical power supply unit required to supply electrical power to the quench water recirculation pump and inline fans. Secondary optional flue gas generators as supplemental CO2 sources, include those shown as gasoline fired internal combustion engine exhaust and/or the natural gas fired refinery cogen unit stack flue gas as selected by opening any of valves V1A through V1E. The combined optional flue gas from optional flue gas sources is fed through stream 1A, where it is combined with stream 1. Dilution air is shown drawn into stream 2, when the in-line fan capacity is higher than the flue gas generation capacity. When the flue gas generation capacity exceeds the in-line fan capacity, the flue gas leaks out of the system through stream 2. When no dilution air is provided, stream 3A has maximized the concentration of CO2.

Since only the quench column is used for cooling, stream 3B to the dry ice sublimizer is zero flow with Valve V3A opened and V3B/V4C closed. The proprietary patent pending quench column receives feed through stream 3B at about 300F, where the hot flue gas flows into the bottom and then up the column through packing that supports intimate contact/mixing with the down flow of recirculated quench water fed and sprayed from the top of the column. The quench water intimately mixes/ directly contacts with the flue gas to absorb heat and condense water from the flue gas. Water is thus produced by condensing out of the flue gas, and heat is neutralized and not accumulated by the competing specific and latent heat processes. The water passes out of and accumulates in a pool at the bottom of the column, and is then pumped via a recirculation pump back to the top of the column. Flue gas exits the top of the column at under 100F as stream 4.

Normal operation is for the flue gas to be sucked through the quench column at either of two capacities with an inline duct fan of either 400 or 1000 CFM as directed by Valves V5B or 5A being opened. The discharge pressurized stream 5 passes into the main header where it then is distributed through sub headers stream 6B, 6C, and 6D through open valves V6A through V6D to then pass to each tree's lateral into each trees stack pipe for crop carbon enrichment with the CO2.

Once lined out with the optimal dispersion and operational parameters (FGSR/FGSC/TSR/TSC/FAC), these values can be used to determine the tonnage of CO2 provided by the unit and to the quench column, and to each tree in Tons CO2 per year (TPY). Additional measurements include weather conditions (AI 1A-D for wind speed, direction, precipitation, humidity), irrigation water to all trees (FI 3), Crop Yield from all Trees (AI 2), Tree Trunk Diameter measurements (AI 3A-I), subsoil moisture at two trees (AI-4,5 A-C, at three subsoil depths of 1,2, and 3 feet), and Manual Leaf Water Potential for tree water utilization efficiency at various times (AI 11 A-D).

Fig 1: Photo of CO2 Generators, Quench Column, In-Line Fans

Fig 2: Photo of Conventional 8" CO2 supply main, 8" branch to 12" stack (3 feet high) to tree.

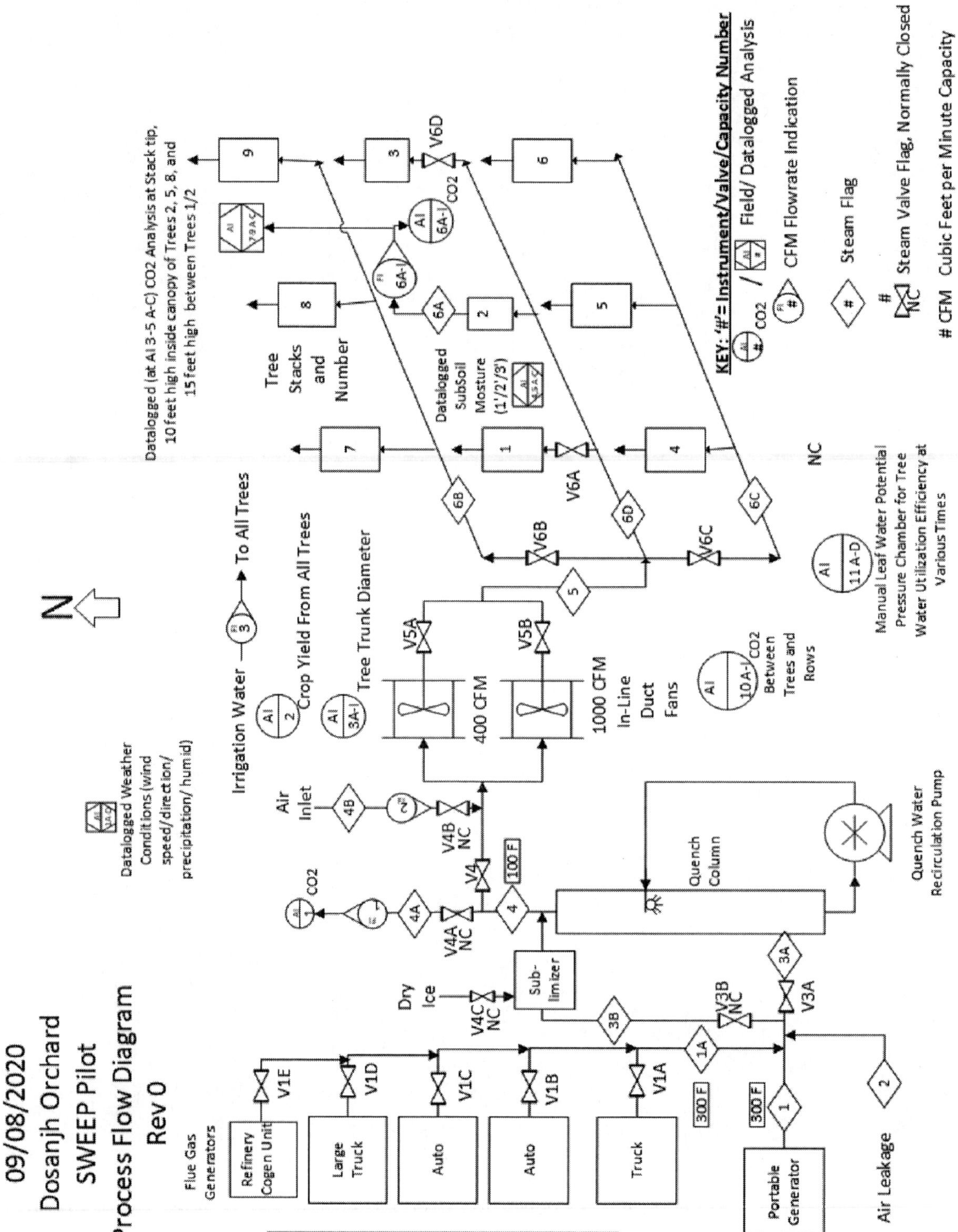

Fig 3: Process Flow Diagram

"...Peace be with you..."

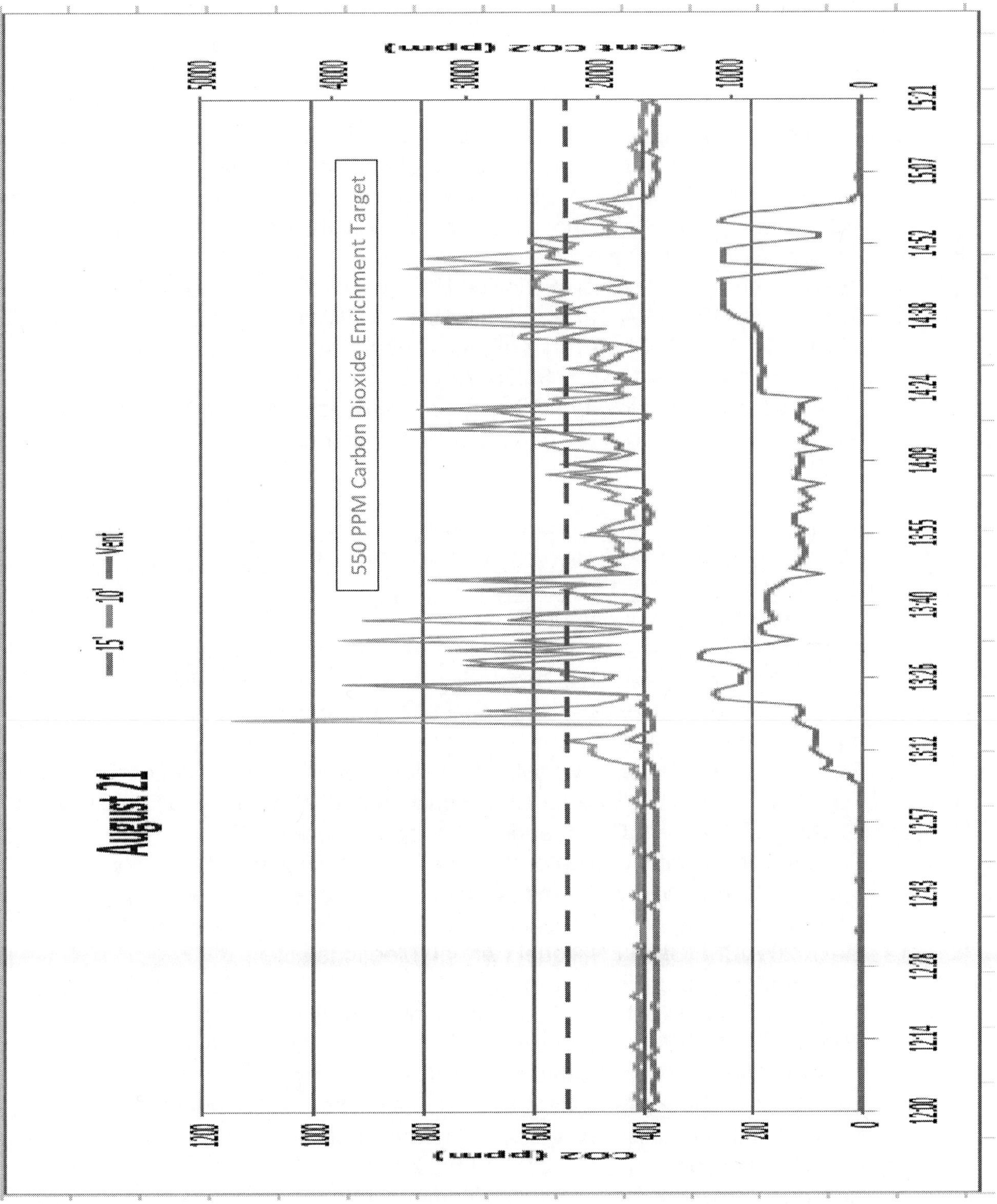

Fig 4: Dosanjh Almond Orchard, CO2 Data, 12PM to 4PM on Friday, August 21, 2020

Accelerating Carbon Neutrality with Sustainable Biomass Cycle

Brian P. Kolodji, PE; Black Swan, LLC; Bakersfield, CA

References:

1. Genesis 2:7
2. De Sassure T: Recherches chemiques sur la vegetation, Paris, 1804. Translated by Wieler A from Chemische Untersuchugen uber die Vegetation, Engelmann, Leipzig, 1890: 22, as sited by Dalrymple DG: A global review of greenhouse food production. US Dept of Agric., Economic Research Service; 1973: 44, as sited by Dr. Kimball in reference 4.
3. "Seventeen Years of Carbon Dioxide Enrichment of Sour Orange Trees: Final Results", by BRUCE A. KIMBALL (USDA), SHERWOOD B. IDSO, STEPHANIE JOHNSON and MATTHIAS C. RILLIG; Global Change Biology (2007) 13, 2171–2183, doi: 10.1111/j.1365-2486.2007.01430.x
4. "Crop responses to elevated CO2 and interactions with H2O, N, and temperature", by Bruce A Kimball (USDA/ Greenleaf Group), Current Opinion in Plant Biology 2016, 31:36–43
5. SunSelect Tehachapi, California 32 Acre Greenhouses and carbon enrichment and credits: https://sempra.mediaroom.com/2018-07-31-SoCalGas-Presents-500-000-Equipment-Incentive-Check-to-SunSelect-Produce-for-New-Energy-Saving-Ultra-Efficient-Cogeneration-Engines and http://sunselect.ca/green-farming/climate-friendly-produce/
6. Results of foundational earlier work for Flue Gas eXtraction and Biosequestration (FGXB™) were presented at the Spring 2018 and from this pilot were was presented at Spring 2019 American Institute of Chemical Engineers (AIChE) national meeting Carbon Management and Sustainability session in Orlando, Florida and New Orleans, Louisiana, the latter of which was attended by Open Top Chamber (OTC)/ Free Air Crop Enrichment (FACE) experts Dr. Kimball of UDSA/The Greenleaf Group and Dr. George Hendrey of CSUNY (retired from Brookhaven National Laboratory). More recent technical presentations have also been made in August 2020 on this technology and operation at the 2020 International Congress on Sustainability Science and Engineering (ICOSSE) and the 2020 AIChE Spring National (Virtual) Meetings
7. **Independent Operations Data Set 1**. Dosanjh Orchard pilot data logged CO2 concentrations at vent, at 10', and 15' above Grade during operations between 7/31 and 8/24/2020 by Dr. Brian Marsh, Director of University of California Cooperative Extension- Kern County (UC Coop-Kern)
8. **Independent Operations Data Set 2.** Notes/ field measurements by Brian Kolodji during commissioning/startup/normal operation/maintenance/ repair of facility during period of 8/5/2020 and 8/25/2020
9. Keeling Curves and atmospheric CO2 measurements and data from Scripps Institute of Oceanography (SIO)and National Oceanographic and Atmospheric Administration (NOAA). https://sioweb.ucsd.edu/programs/keelingcurve/

591132 Jetty Bog Recovery for Rich LNG

Abstract:

A considerable amount of BOG is generated during the LNG tanker loading for the base load LNG plant. Jetty BOG from older generation baseload LNG plants has not been recovered, since the jetty BOG generation is intermittent and to recover the jetty BOG is costly. However, due to recent concern about the global warming by CO2 emission from the flare stack, the jetty BOG recovery has become significant issue nowadays. Generally speaking, the BOG is leaner compared with the baseload feed gas, therefore to recovery the BOG as feed gas, there will be some effect to the liquefaction performance.

A previous presentation of recycling the Jetty BOG back to the liquefaction train showed minor improvements, mostly due to the lean nature of the overall feed gas. These results were performed by using CHEMCAD by Chemstation to simulate the overall process. This presentation will build off of previous work to look at using rich LNG BOG and recycling back to the front of the LNG train.

590324 Performance Analysis of LNG Liquefaction Process Integrated with Allam Cycle Centered Power Plant

Abstract:

Liquefied Natural Gas (LNG) is an increasingly important clean energy source, which has been widely produced, transported, and consumed all over the world. Particularly in the US, the ambulant availability of natural gas enables many LNG liquefaction projects to be built in the following decade. In the meantime, the increasing trend to utilize the natural gas to generate power becomes an important element to monetarize the natural gas in the U.S. In this paper, an LNG liquefaction process and an Allam cycle centered power plant has been integrated for study. The power plant subsystem is the natural gas empowered complex, which includes a power plant driven by the Allam cycle to generate net power and pure CO_2 for commercial utilization; an air separation unit (ASU) to supply the pure oxygen to the power plant and the pure nitrogen as feed stock; as well as an ammonia plant to utilize the nitrogen to produce ammonia. The LNG liquefaction system is the propane pre-cooled mixed refrigerant process (C3MR), which involves pre-cooling of natural gas using propane refrigeration loop followed by sub-cooling and liquefaction in the second loop using a mixed refrigerant composed of methane, ethane, and propane.

Note that the nitrogen gas stream is a by-product from the ASU unit of the power plant subsystem, which is at a temperature of -25^0C and can be used in the C3MR process for help pre-cooling operation in advance, and then be directed to the ammonia plant as the feed stock in order to save energy. In this paper, we have explored various opportunities of utilizing this nitrogen gas stream for helping the C3MR pre-cooling process. This study targets on maximizing the energy efficiency and minimizing the cost of production simultaneously. This integrated new process has been virtual demonstrated via simulation, which is economically sound, technological viable, and environmentally benign due to the comprehensively monetarized cheap natural gas resource to accomplish an integrated power-chemical-LNG process via heat, mass, and work integrations.

Vapor Dispersion: How Safe is your Safety Factor?

Jessica M. Morris, Ph.D.
Ryan Hart, Ph.D., P.E.
Delmar "Trey" Morrison, Ph.D., P.E., CPSP
Exponent, Inc.
4580 Weaver Parkway, Suite 100,
Warrenville, Illinois, USA 60555
jmorris@exponent.com

Prepared for Presentation at
American Institute of Chemical Engineers
2020 Spring Meeting and 16th Global Congress on Process Safety
Virtual Conference
August 17 - 21, 2020

Vapor Dispersion: How Safe is your Safety Factor?

Jessica M. Morris, Ph.D.
Ryan Hart, Ph.D., P.E.
Delmar "Trey" Morrison, Ph.D., P.E., CPSP
Exponent, Inc.
4580 Weaver Parkway, Suite 100,
Warrenville, Illinois, USA 60555
jmorris@exponent.com

Keywords: Manage Risks, Safety & Health and Transport Phenomena

Abstract

Risk analysis, consequence analysis, building siting, demonstrating compliance with regulatory permitting criteria – these activities all rely heavily upon mathematical modeling of physical phenomena and hazardous vapor dispersion. The common safety factor or rule of thumb, especially as applied in permitting of LNG facilities through the Pipelines and Hazardous Materials Safety Administration (PHMSA), applies a value of 2 for flammable vapor dispersion. The hazard threshold is often modeled as ½ of the lower flammable limit (LFL). But who came up with the safety factor and how useful is this factor of 2? In the case of the PHMSA Model Evaluation Protocol, this was loosely based upon standard models' concurrence with a set of large scale field test results. In this manuscript, we will briefly introduce those results but use a different and new data set to evaluate the significance of this safety factor selection for flammable gas dispersion. Downwind concentration time histories for multiple propane releases in a controlled wind tunnel environment will be analyzed both individually and as an ensemble average resulting in an experimentally significant safety factor for comparison to the existing guidance.

1 Introduction

Hazardous chemicals are often found in large quantities at industrial facilities due to processing and storage demands. With a larger quantity of chemicals in a single location, it is paramount to ensure that if these chemicals are accidentally released, the personnel and environment are kept safe. Release scenarios and their consequences can be modeled in risk analysis, consequence analysis, building siting, permitting applications, and emergency response planning. An example of a simplified model is shown below that incorporates a safety factor of two.

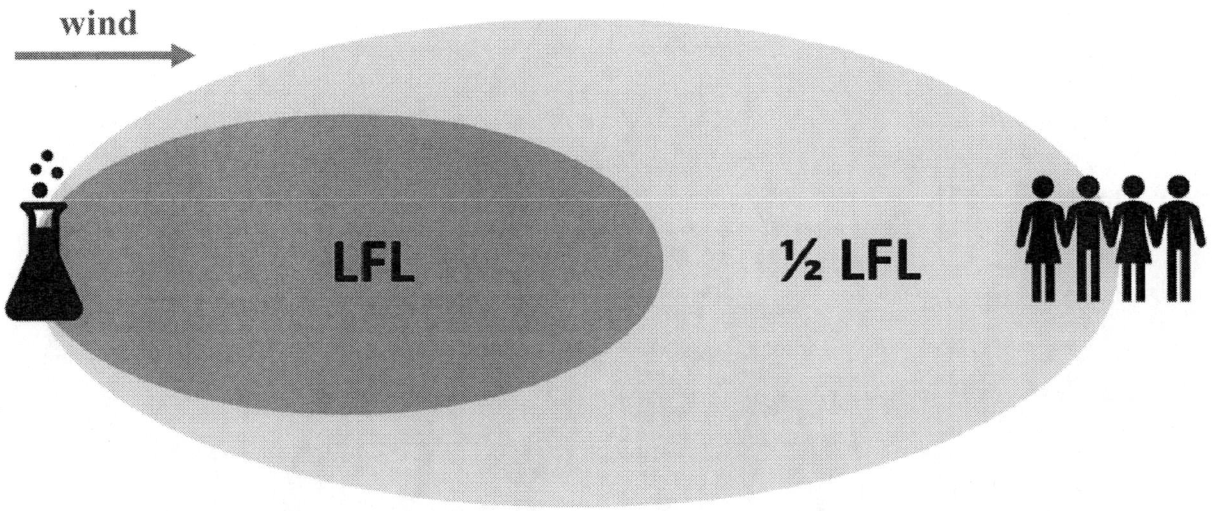

Figure 1. Sample flammable release modeled to the LFL and the ½ LFL to determine an appropriate distance to downwind receptors.

The modeling of releases typically includes a safety factor, normally with a value of 2, to determine what distance is appropriate to keep personnel, property, and the environment safe. The safety factor is intended to compensate for potential under-estimation of real world concentrations using mathematical models that provide discrete solutions. Although a safety factor of 2 results in a more conservative effect distance estimate relative to no safety factor, in a real release, gasses are not uniformly distributed throughout a cloud. Wind currents, instability, and turbulence, among other phenomena, can result in pockets of gas with higher or lower concentration than would be calculated from a model. Application of an appropriate safety factor is critical in minimizing the risk posed by accidental releases.

1.1 Safety Factor

When a contaminant is released into the atmosphere, it mixes with the surrounding air resulting in a decreased concentration as a function of time and distance. But the mixing of gasses is not a uniform phenomenon – at a specified measurement point, the observed concentrations will fluctuate with time for a constant release source. An ensemble average of time-varying concentration data for multiple test replicates represents a more uniform distribution of concentration and is the average concentration that is commonly used in validation testing of models resulting in the predicted or benchmark concentration. A comparison of the ensemble average concentration versus the fluctuations in a single test is shown below. The figure shows concentration data for propane as a fraction of the source concentration at a specific measurement location downwind for the duration of the release. The blue data points represent the measured concentration at the observation point versus time during the test. The red data points represent the ensemble average of the concentration measured over multiple tests. For example, the ensemble average value at 40 seconds represents the average of the concentration at the measurement point at 40 seconds over all tests. In this way, the ensemble average smooths out the random fluctuations in concentration from test to test and time to time.

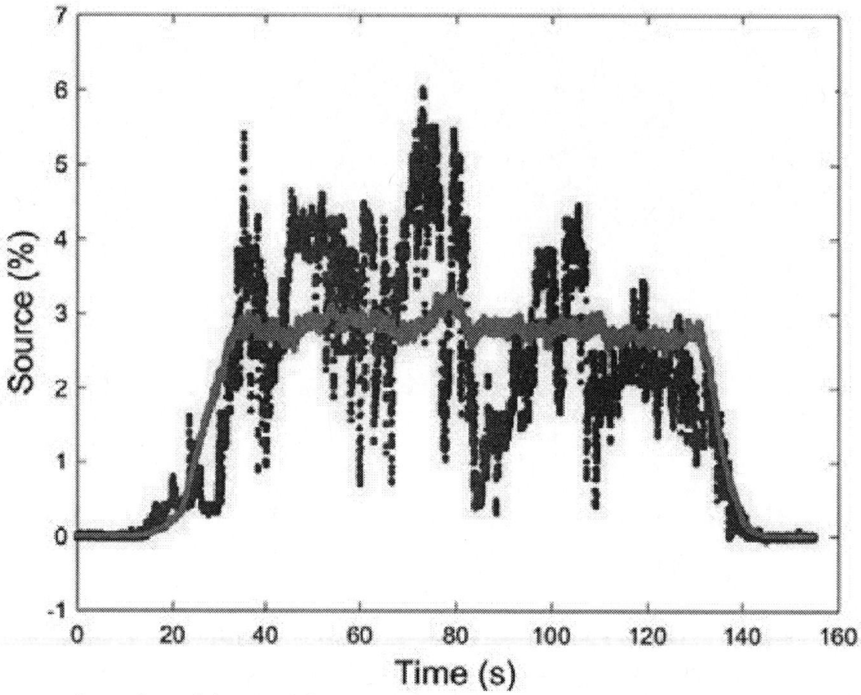

Figure 2. Concentration time history demonstrating the fluctuations in concentration from a single test (blue) vs. ensemble average (red).

As observed in the figure, the single test concentration (blue data points) can vary significantly although the release source and wind source are constant. When the concentration is much larger (or much smaller) than the averaged ensemble concentration, this represents localized pockets of gas existing at smaller or larger concentrations than would be calculated in a discrete mathematical model. Due to real concentration fluctuations, a safety factor is commonly applied to address the uncertainty in the actual value at a specific point at a specific time. A safety factor of 2 can be practically applied by using the distance to the ½ lower flammable limit (½-LFL) as opposed to the LFL as required in permitting, building siting, emergency planning, and other applications.

1.2 Validation of safety factor with previous data

Field tests are primarily single tests due to the uncontrolled variation in wind speed, direction, solar insolation, and overall weather. The safety factor to incorporate concentration fluctuations can therefore best be validated with a large sample of tests under uniform conditions. Wind tunnels allow for multiple controlled tests with consistent wind speed, wind direction, heat transfer (solar insolation), precipitation, and other variables. Due to this repeatability and the controlled parameters in a wind tunnel, concentration time histories can be measured multiple times under the same parameters allowing for random concentration fluctuations to be analyzed.

In this paper, concentration fluctuations from an experimental program performed in an ultra-low speed wind tunnel will be analyzed to determine if the safety factor of 2 is appropriate.

2 Experimental Program

The original experimental program took place in an ultra-low speed wind tunnel with controlled parameters to determine the effects of along-wind dispersion as a function of wind speed, downwind distance, and release duration. This data set[1] will be further analyzed to determine an appropriate safety factor due to concentration fluctuations that were measured in the original experimental program.

2.1 Facility and Equipment

Testing was performed in an ultra-low speed wind tunnel at the Chemical Hazards Research Center (CHRC). The wind speeds are less than 2 m/s, measured by a 3D Laser Doppler Velocimeter (LDV). Through a modified Irwin spire and surface roughness arrangement, a 40cm constant stress layer was achieved ensuring the tests are well within this layer to simulate the atmosphere. Concentration time histories were measured with a fast Flame Ionization Detector (fastFID) with essentially no time-delay. The source gas was released through a novel area source fabricated for steady-state finite duration releases. Additional descriptions of the facility and test program can be found in previous publications.[1, 2]

2.1.1 Ultra-Low Speed Wind Tunnel

The wind tunnel at the Chemical Hazards Research Center has a working section of 24 m x 6.1 m x 2.1 m in length, width, and height respectively. The boundary layer is established through a height of 60 cm and a constant stress layer of 40 cm.[3] Wind speeds from 0.34 m/s to 1.1 m/s are produced through the "push through" wind tunnel resulting in the classification of ultra-low speed and for scalability for atmospheric conditions.[4] The wind tunnel is a closed loop wind tunnel resulting in no disruption of flow and a more controlled environment.

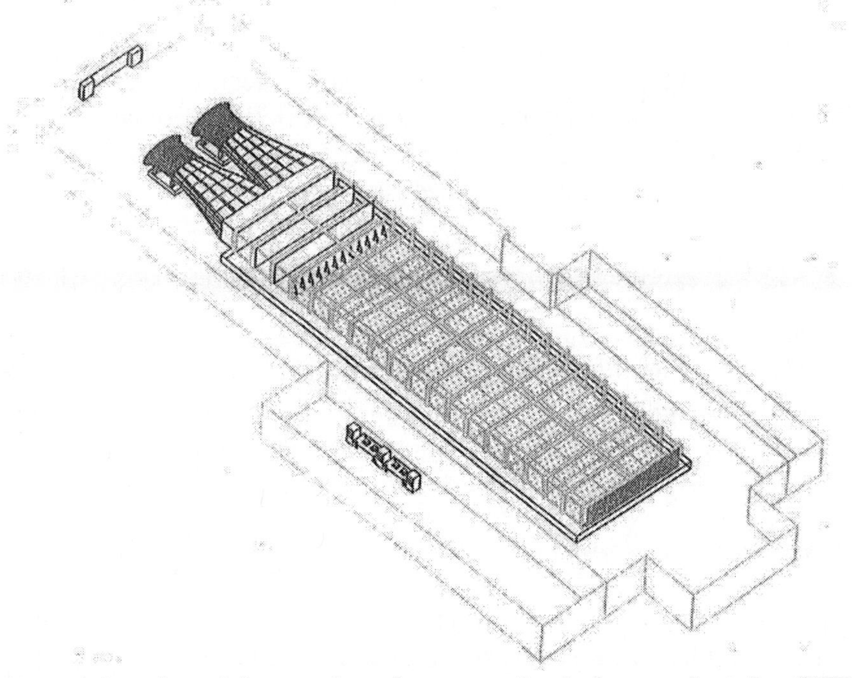

Figure 3. Floor plan of the closed-loop, ultra-low speed wind tunnel at the CHRC.[5]

2.1.2 Concentration Measurements

The concentration time histories were all recorded through a Cambustion HFR-400 Atmospheric fast Flame Ionization Detector (fastFID). The fastFID is a non-specific aspirated probe used to detect propane in the experimental program. A sample rate of 100 Hz was obtained. Additional information on the specifics and data collection parameters can be found in previous publications. [1, 2]

2.2 Overall Test Matrix

The experimental program was established to study and determine an accurate along-wind dispersion coefficient. To accomplish this, an extensive test matrix was performed encompassing multiple wind speeds, release durations, and downwind distances resulting in thousands of individual tests performed. The steady-state data from the 120-second releases in this experimental program were analyzed for concentration fluctuations to determine an appropriate safety factor. A summary of the test parameters that achieved steady-state at 120-second finite-duration releases are shown below in Table 1.

Table 1. Test conditions from 120-second finite-duration releases.

Test	Wind Speed of wind tunnel at top of boundary layer (m/s)	Wind Speed of wind tunnel at 1cm (m/s)	Distance Downwind from Area Source (ft)	Tests Performed at specified conditions
A	0.71	0.224	4	65
B	0.71	0.224	8	100
C	0.71	0.224	12	67
D	0.35	0.102	2	65
E	0.35	0.102	4	65
F	0.35	0.102	8	100

Variability in the total tests performed was a function of concentration fluctuations observed in the ensemble. If the ensemble of all tests had visual fluctuations in concentration after the initial 65 tests were recorded, additional tests were performed. Additional analysis of the concentration fluctuations of these tests was performed to determine an appropriate safety factor.

3 Analysis of Data

The preliminary analysis of the concentration fluctuations starts with the determination of steady-state. Only the steady-state periods of the releases were analyzed to determine an appropriate safety factor. The maximum concentration observed in each of the time histories was extracted from the data to determine the upper limit of concentration that was observed for each of the set of tests. A basic statistical analysis was also performed for thoroughness of analysis.

3.1 Steady-State Concentration

The steady-state concentration was determined for each of the scenarios. Accurate determination of the steady-state concentration was determined with 10-minute release durations at the respective downwind location and windspeed. This 10-minute averaged concentration, or "peak' observed concentration, was applied to the 120-second releases as the steady-state concentration at the respective downwind distance and windspeed. To ensure the steady-state or "peak" concentration was accurately quantified, a variation of <0.1% of a moving average was applied across the ensembled tests.[1] If the moving average fell outside of this variation, steady-state was not deemed reached. All 120-seconds releases in the experimental program achieved steady-state or "peak" concentration. A table with the time of peak arrival (TOPa) and time of peak departure (TOPd) with the total duration steady state was achieved is shown below.

Table 2. Time of peak arrival and departure for each test with duration of steady state.[1]

Scenario	Wind Speed (m/s)	Distance (ft)	TOPa (s)	TOPd (s)	Duration of Steady-State
A	0.71	4	42.45	123.9	81.45
B	0.71	8	45.43	128.4	82.97
C	0.71	12	52.19	131.5	79.31
D	0.35	2	58.50	126.4	67.90
E	0.35	4	66.20	128.0	61.80
F	0.35	8	74.47	133.3	58.83

3.2 Maximum Observed Concentration

The maximum observed concentration for each of the ensembled concentration time histories as a function of time was extracted for a visual representation of the data. The plots show that the maximum observed concentrations (red, Figure 2) are significantly larger than the ensemble averaged concentrations (black, Figure 2). The magnitude of the difference in maximum and ensemble concentrations will be discussed in subsequent sections.

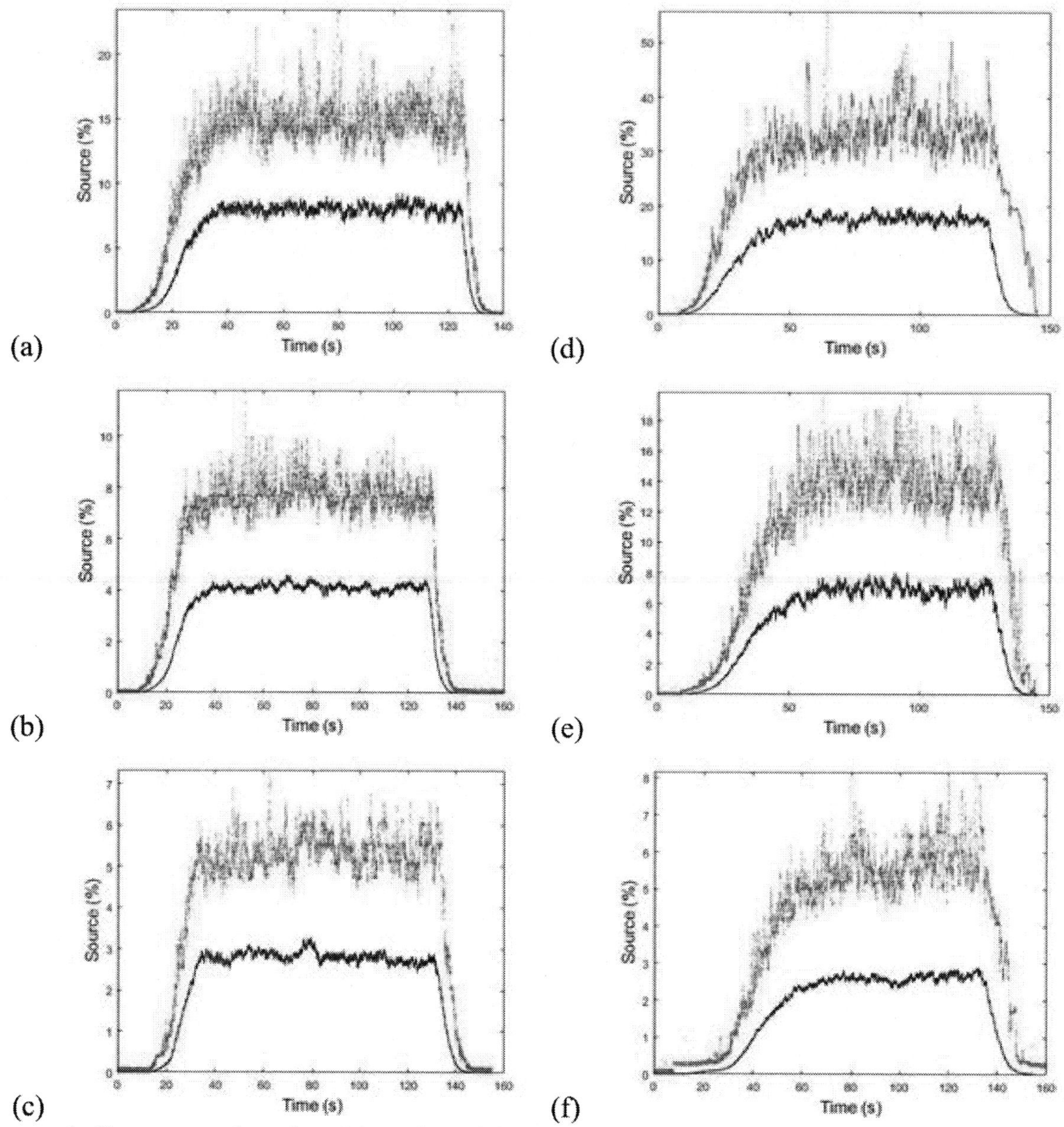

Figure 4. Concentration time histories with ensembled concentrations (black) and maximum concentrations (red) at downwind distances of 4 ft (a), 8 ft (b), and 12 ft (c) for 0.71 m/s and 2 ft (d), 4 ft (e), and 8 ft (f) for 0.35 m/s.

3.3 *Statistical Analysis*

A basic statistical analysis was performed for each set of scenarios. The ensembled mean, maximum, minimum, and standard deviation averages for the steady-state periods are shown below in Table 3. The values are a function of the percentage of the source gas observed, which was 1% propane in air for all release scenarios.[1]

Table 3. Ensembled Mean, Maximum, Minimum, and Standard Deviation averages for the steady-state data.

				Percentage of Source Gas			
Scenario	Wind Speed (m/s)	Dist. (ft)	Tests	Ensembled Mean Avg.	Ensembled Max Avg.	Ensembled Min Avg.	Ensembled STD Avg.
A	0.71	4	65	8.09	15.1	1.31	3.01
B	0.71	8	100	4.13	7.78	0.640	1.48
C	0.71	12	67	2.81	5.40	0.136	1.24
D	0.35	2	65	17.8	34.3	2.76	6.81
E	0.35	4	65	6.98	14.2	0.945	2.88
F	0.35	8	100	2.63	5.74	-0.0283	1.14

From the values observed in the table, the maximum concentration is not always twice (2x) the average concentration, depending on the time and downwind distance. An appropriate safety factor was estimated through additional analysis. Visual representations of the statistical data for Scenario B with 100 recorded tests is shown below.

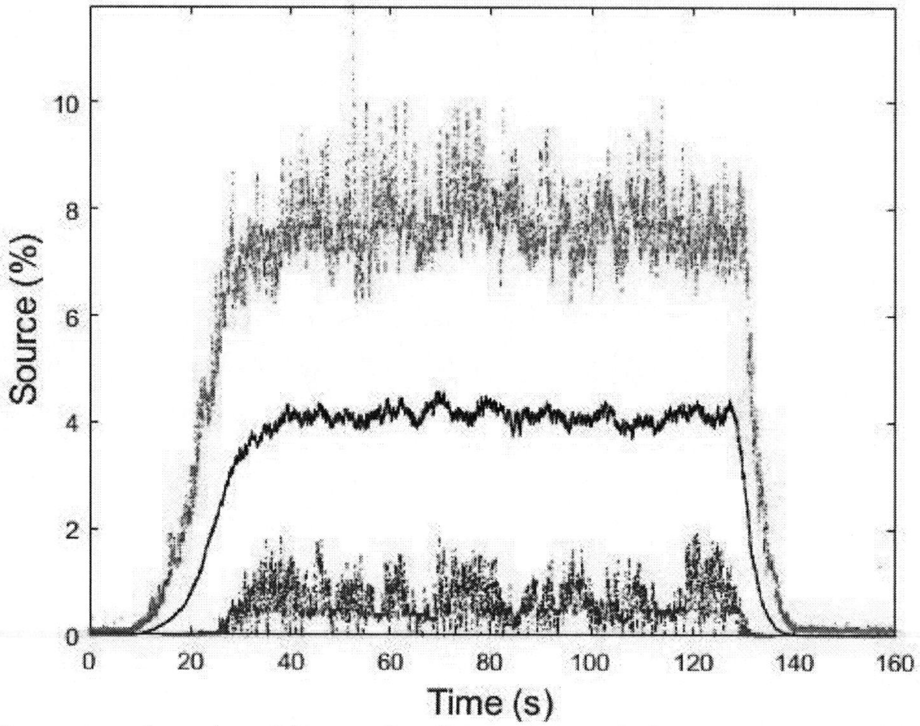

Figure 5. Concentration time history for Scenario B (8 ft downwind and 0.71 m/s) with ensembled concentrations (black), maximum concentrations (red), and minimum concentrations (blue).

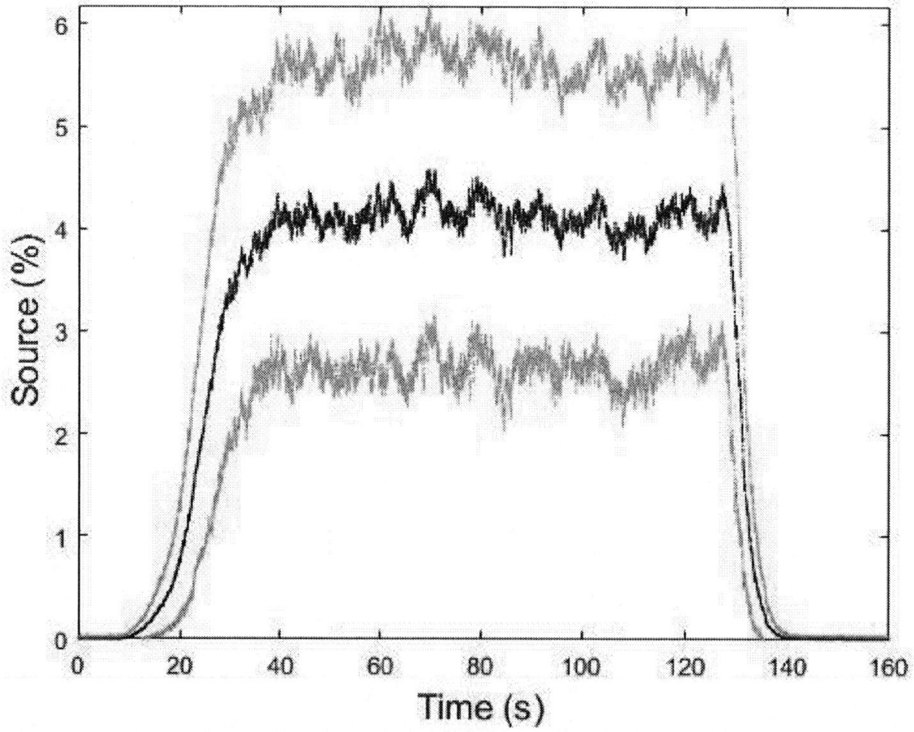

Figure 6. Concentration time history for Scenario B (8 ft downwind and 0.71 m/s) with ensembled concentrations (black) and standard deviations (orange).

4 Results and Discussion

Two different methods for determining a safety factor was considered. In the first approach, the safety factor was determined as a function of the individual observed concentrations at each steady-state sample time using a statistical percentile method. The second safety factor was determined as a function of the maximum observed concentrations at each steady-state sample time for a more conservative safety factor as a function of absolute data. The two methods with results will be discussed in further detail below.

4.1 Safety Factor for All Data as function of Percentile

Each scenario had multiple tests obtained under the same parameters. Concentrations were measured every 0.01 seconds for each test. Ensembles were performed by taking the average of all tests at the individual sample time. This ensemble method encompasses a large range of data for each sample time that can be further analyzed.

A safety factor was determined for each scenario as a function of percentiles for each set of tests for the steady-state data. For a specified percentile, there is a concentration in which all observed data falls below this value. A 75% percentile would give the observed concentration where 75% of the measured values are less than the reported value and 25% of the measured values are greater than the reported value. This 75% percentile is shown visually in Figure 7 above with the green line.

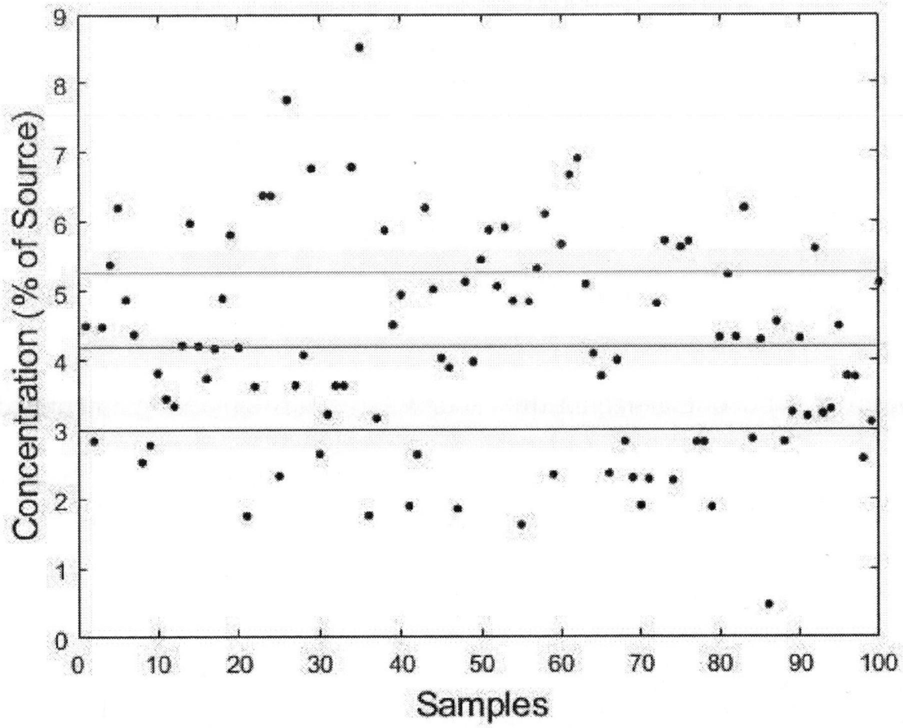

Figure 7. All concentrations measured for 100 individual tests for Scenario B (8ft downwind and 0.71 m/s) at 90.00 seconds. Individual measured concentrations (black) from each test with lines breaking the data into 25% (blue), 50% (red), and 75% (green) of values encompassed below the lines (percentiles).

The safety factor was estimated for the specified percentile with the following equation:

$$Safety\ Factor = \frac{Percentile\ Concentration}{Mean\ Concentraiton} \tag{1}$$

A range of safety factors are shown in Table 4 below for percentiles from 50% to 99.9%.

Table 4. Safety Factor needed to achieve the percentiles.

Scenario	Wind Speed (m/s)	Distance (ft)	Safety Factor				
			50%	75%	95%	99%	99.9%
A	0.71	4	1.00	1.26	1.60	1.84	1.86
B	0.71	8	1.01	1.24	1.59	1.83	1.88
C	0.71	12	1.03	1.33	1.68	1.90	1.92
D	0.35	2	1.01	1.26	1.61	1.90	1.93
E	0.35	4	1.00	1.27	1.70	2.01	2.03
F	0.35	8	0.95	1.30	1.76	2.09	2.19

From these safety factor estimates, using the 50% percentile results in a safety factor close to 1, which is expected since this approximates the average concentration. For a safety factor than encompasses 99.9% of observed data, the safety factor is around 2. This aligns with the commonly used safety factor of 2 and captures the statistically significant portion of the data while neglecting outliers. The different wind speeds and downwind distances resulted in slightly different safety factors. For slower wind speeds, a larger safety factor is needed to encompass the concentration fluctuations observed at the specified percentile. At increased downwind distance, a larger the safety factor is needed to encompass the concentration fluctuations observed at the specified percentile.

4.2 Upper Limit Safety Factor for Maximum Observed Data

While the percentile method statistically shows the observed data points at each sample time that falls within the specified percentile, real world applications may not be as statistically simple. If a flammable cloud is traveling through the atmosphere and the concentration exceeds the LFL for any duration of time yet encounters an ignition source, a flash fire or explosion may occur. Pockets of gas with higher concentration can provide catastrophic results[5] and are observed with the maximum observed concentrations in this experimental program. This phenomena is visually represented in the figures below with the mean and max concentrations plotted as a function of time with the safety factor of 2.

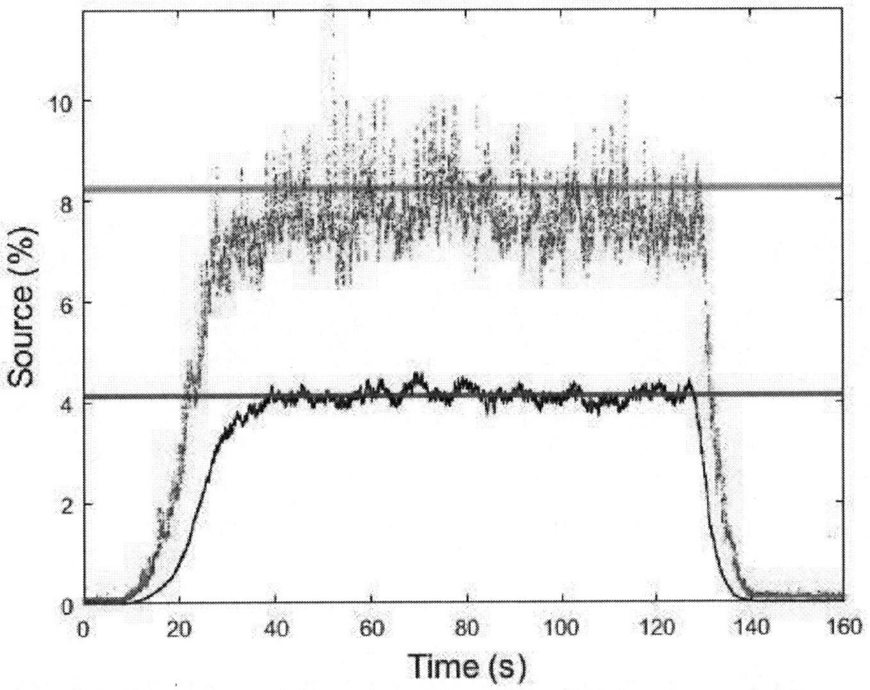

Figure 8. Concentration time history with ensembled concentrations (black), mean concentration (blue), maximum concentrations (red), safety factor of 2 (gray) for Scenario B (8 ft downwind and 0.71 m/s).

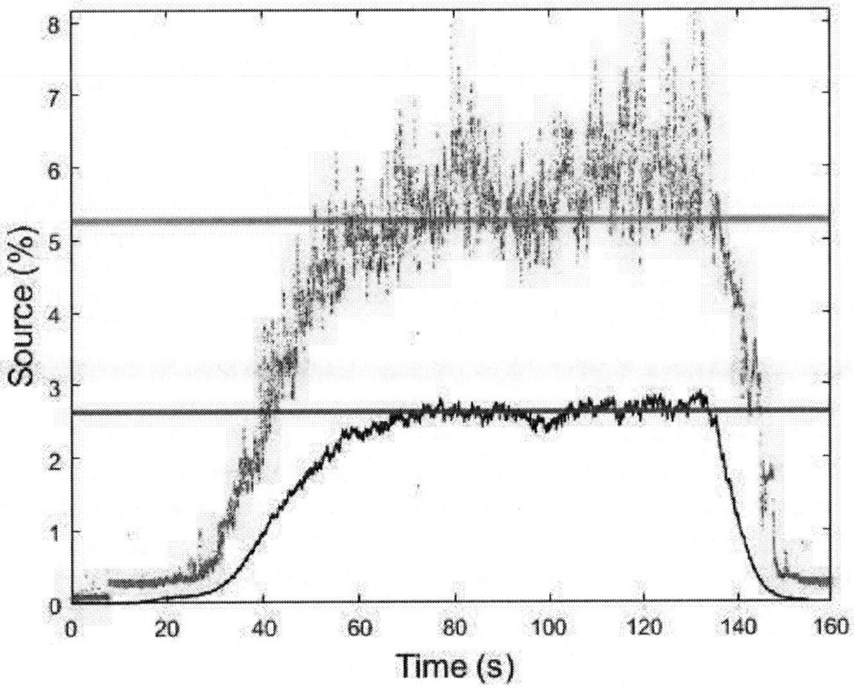

Figure 9. Concentration time history with ensembled concentrations (black), mean concentration (blue), maximum concentrations (red), safety factor of 2 (gray) for Scenario F (8 ft downwind and 0.35 m/s).

While a safety factor of 2 sounds conservative, the visual representation of the blue line shows that there are many data points where the maximum observed concentration in an individual test exceeded this applied safety factor. A percentage of the data in which the steady-state maximum observed concentration exceeded the safety factor of 2 is shown in the table below.

Table 5. Percentage of the maximum steady-state observed concentration exceeds the safety factor of 2.

Scenario	U (m/s)	Distance (ft)	Maximum observed concentration exceeds 2 x predicted concentration
A	0.71	4	20.0%
B	0.71	8	18.8%
C	0.71	12	24.0%
D	0.35	2	29.8%
E	0.35	4	54.2%
F	0.35	8	76.6%

A more conservative option to minimize the risk that the calculated distance is would be to set the safety factor according to the maximum value observed as an upper limit safety factor. Upper limit safety factors were determined for each set of scenarios to incorporate 95%, 99%, and 99.9% of the time where the maximum observed concentrations are below the maximum safety factor multiplied by the ensembled mean concentration. These upper limit safety factors are shown in the table below.

Table 6. Upper Limit Safety Factor needed to encompass the maximum steady-state observed concentrations as a function of time for each scenario.

Scenario	U (m/s)	Distance (ft)	95%	99%	99.9%
A	0.71	4	2.18	2.45	2.76
B	0.71	8	2.16	2.32	2.56
C	0.71	12	2.17	2.28	2.49
D	0.35	2	2.38	2.62	2.99
E	0.35	4	2.45	2.63	2.74
F	0.35	8	2.59	2.84	3.01

The upper limit safety factors provide a more conservative option to incorporate the potential for the maximum observed concentration that may be achieved at the sample time. Similar to the safety factor encompassing all data, the upper limit safety factor as a function of only maximum observed data, shows to be a function of the wind speed and downwind distance. Visual representations of these upper limit safety factors are shown below.

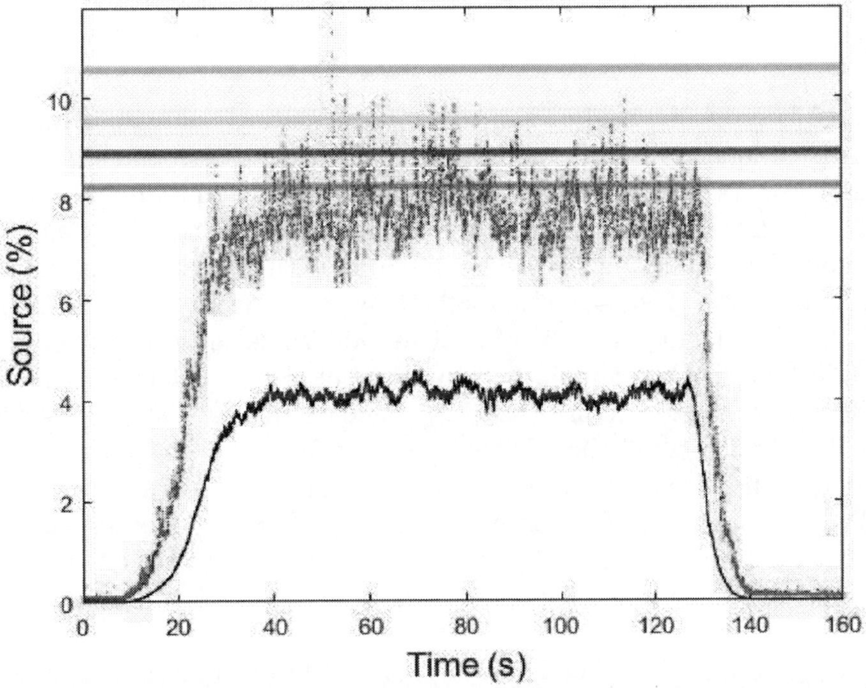

Figure 10. Concentration time history with ensembled concentrations (black), maximum concentrations (red), safety factor of 2 (gray), 95% (blue), 99% (green), and 99.9% (orange) of maximum observed concentration for Scenario B (8 ft downwind and 0.71 m/s).

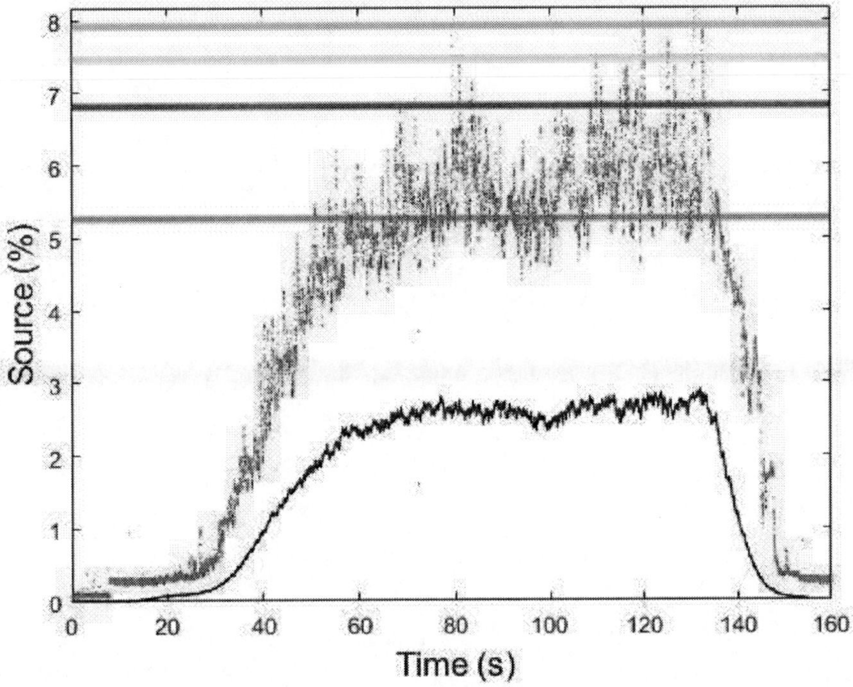

Figure 11. Concentration time history with ensembled concentrations (black), maximum concentrations (red), safety factor of 2 (gray), 95% (blue), 99% (green), and 99.9% (orange) of maximum observed concentration for Scenario F (8 ft downwind and 0.35 m/s).

5 Summary

A safety factor of 2 is typically used in industry for calculating flammable and toxic effects but has not been validated against a large repeatable data set due to repeatability of field tests and limited data in ultra-low speed wind. The experimental program at the Chemical Hazards Research Center provided thousands of tests, where the steady-state releases were analyzed further to determine an appropriate safety factor as a function of time that appropriately incorporated fluctuations in concentration. Two different methods were performed to determine an appropriate safety factor. 1) A safety factor as a function of the individual observed concentrations at each steady-state sample time using a simple percentile method was determined. 2) An upper limit safety factor a function of the maximum observed concentrations at each steady-state sample time was determined. The first method showed that a safety factor of 2 was able to capture 95% of concentration fluctuations for all scenarios. For larger wind speeds, a safety factor of 2 was able to capture 99.9% of concentration fluctuations at each sample time. For slower wind speeds and increased downwind distance, a slightly larger safety factor would be needed to capture 99% and 99.9% of concentration fluctuations. The upper limit safety factor method showed that in order to incorporate all maximum observed concentration fluctuations, an upper limit safety factor of 2 is insufficient. The upper limit safety factor similarly showed trends being a function of wind speed and downwind distance.

6 References

[1] J.M. Morris. "Experimentation and Modeling of the Effects of Along-Wind Dispersion on Cloud Characteristics of Finite-Duration Contaminant Releases in the Atmosphere," PhD Dissertation, University of Arkansas, 2018.

[2] J.M. Morris and T.O. Spicer III. "Improved Modeling of Continuous Gas Releases of Finite Duration." 15th Global Congress on Process Safety, New Orleans, LA. 2019.

[3] J.M. Morris, T.O. Spicer III, and J.A. Arthur. "Three-Dimension Flow Structure of a Simulated Atmospheric Boundary Layer using an Irwin Spire-Roughness Technique." Submitted to *Boundary Layer Meteorology*. 2019.

[4] Havens, J., and T. Spicer. "Vapor dispersion and thermal hazard modelling." *Final topical report to Gas Technology Institute under sub-contract K* 100029184. 2006.

[5] Walker, H., and J. Havens. "Comparison of hot wire and laser Doppler anemometry measurements of the turbulence structure of an ultra-low speed wind tunnel boundary layer." *Data report, Chemical Hazards Research Center,* University of Arkansas, Fayetteville, AR 72701 (1998).

[6] R. A. Ogle. "Explosion Hazard alaysis for an Enclosure Partially Filled with a Flammable Gas," *Process Safety Progress*, pp. 170-177. Fall 1999.

Application of Safety Instrumented Systems in Liquified Natural Gas (LNG) Processes

Denise Chastain-Knight
Exida Consulting, LLC
Sellersville, PA
dchastainknight@exida.com

Patrick O'Brien
Exida Consulting, LLC
Sellersville, PA
PObrien@exida.com

Prepared for Presentation at
American Institute of Chemical Engineers
2020 Spring Meeting and 16th Global Congress on Process Safety
Houston, TX
March 30 – April 1, 2020

Application of Safety Instrumented Systems in Liquified Natural Gas (LNG) Processes

Denise Chastain-Knight
Exida Consulting, LLC
Sellersville, PA
dchastainknight@exida.com

Patrick O'Brien
Exida Consulting, LLC
Sellersville, PA
PObrien@exida.com

Keywords: Safety Instrumentation Systems, Process Design & Development

Abstract

At first glance, one might assume that Liquified Natural Gas (LNG) processing is simply a compression and cooling process. Deeper observation reveals unique process challenges of flammable and explosible materials, high-pressure rotating equipment, and cryogenic conditions. LNG facilities are typically developed by connecting highly specialized process units designed and constructed by technology providers. Establishing a consistent and effective Safety Instrumented System (SIS) philosophy across the entire facility requires careful thought and specification to align all technology providers to the same page. This paper will highlight some of the project management and design concerns when working with packaged equipment through a study of some specific Safety Instrumented Function (SIF) examples.

1 Introduction

Natural gas (NG) is a flammable material comprised of light hydrocarbons (methane, propane, butane) non-condensable (CO_2) and some trace elements (H_20, H_2S, mercury). It is desirable to process NG into the liquified natural gas (LNG) because it is easily stored at near atmospheric pressure at very low temperature and can be transported long distances. The liquifaction process will reduce the volume to 1/600th of the gas form [1] however, the energy potential is also concentrated raising the overall risk per unit volume. The boiling point (-306°F) and flash point (-256°F) [2] are extremely low and if containment is lost LNG will readily vaporize forming a flammable and potentially explosive vapor cloud with asphyxiant and cryogenic hazards.

The processing of NG to LNG includes pretreatment to remove the undesired components (H_2S, CO_2, water, mercury) and then a cryogenic refrigeration process to supercool the gas to a liquified state. Safety Instrumented Systems (SIS) are an important layer of protection in the LNG process. The system is called on to identify conditions that could lead to a loss of containment, and act to bring the process to a safe state. For the SIS to be successful, designers must be aware of the special considerations of the process when specifying sensor and final elements, performing design verification, and performing site integration of packaged systems.

This paper will provide insight on design issues for SIS in LNG applications including industry lessons learned, considerations for integrating package systems with vendor supplied Basic Process Control System (BPCS) and SIS into an overall site system. Additionally, specific considerations for the selection of sensors and final elements in Safety Instrumented Functions (SIF) considering cryogenic applications is provided. Finally, multiple Safety Integrity Level (SIL) verification calculation methodologies will be compared to illustrate the impact of assumptions and variable assignment on the calculation outcome, and the impact of the operations and maintenance phase on the SIL verification results is explored.

2 LNG Industry Lessons Learned

A SIS is a collection of SIFs each designed to protect against a specific hazard scenario. Assuring that the SIS is designed to provide reliable protection begins with an understanding of the hazards specific to the service and the hazard scenarios that can develop. To gain this understanding, we look to industry experience for lessons learned. Table 1 is a summary of some significant events experience in LNG history.

Table 1. LNG Industry Event Summary

Event	**Key Details**	**Lessons Potentially Impacting SIS Design**
Feyzin Refinery Feyzin, France 1966[3]	• 15-18 fatalities, 80 injured • Loss of containment due to valve freezing & mechanical failure during manual draining operation • Vapor cloud traveled to an ignition source; flash fire propagated back to the leak source and continued to burn • Sphere overheated and BLEVE occurred	• Valve plugged by ice or hydrate • No remotely actuated emergency isolation • No combustible gas detection • No remotely actuated depressurization • Inadequate water spray to cool sphere – metal fatigue on unwetted portion of sphere

Event	Key Details	Lessons Potentially Impacting SIS Design
Pemex Mexico City November 1984[4]	• 600 fatalities, 7000 injured, 200,000 evacuated • Storage and Distribution Terminal destroyed • LPG Leak into a diked area; vapor cloud over-spilled containment, was ignited, flame front accelerated back to source of leak causing further rupture and nine additional explosions and BLEVE occurred.	• Scenario cause believed to be an overfill and/or overpressure of interconnecting piping • Inadequate or ineffective gas detection system • Late initiation of SIS therefore it was ineffective. • LPG flow to terminal continued for 1 hr after event • Active fire protection disabled by blast
ESSO and BHP[1] Longford, Victoria Australia 1998[5]	• 2 Fatalities, multiple injuries • All production from facility ceased prompting an order to cease all but essential gas usage in the region supplied by the facility. • Loss of hot oil flow to absorbers and subsequent cooling led to low temperature causing steel embrittlement. Vessel ruptured when hot oil was re-introduced. • Hydrocarbon vapor release ignition; explosion and fire • Initial explosion caused secondary releases and additional explosions occurred.	• Process override created high-level condition that led to hot oil pump trip. No written procedures or training to guide personnel on proper response. • Level could not be monitored once the instrument range was exceeded and level history was not recorded. • ESD was not designed to isolate flammable inventory.
Sonatrach Skikda Algeria January 2004[6]	• 27 fatalities, 56 injured • Major LNG leak drawn into boiler firebox causing the initial explosion • Boiler explosion ignited vapor cloud causing secondary consequences	• Detect leaks early while they are small • Isolate systems to minimize volume of loss of containment

[1] Esso Australian Resources Ltd. And BHP Petroleum (Bass Strait) Pty Ltd.

Event	Key Details	Lessons Potentially Impacting SIS Design
Enterprise Products Pascagoula, MS June 2016[7]	• No fatalities, voluntary evacuations • $17.5 million expense and non-cash loss • Release of methane, ethane, propane (LNG, MR[2]) when BAHX[3] failed ignited, initiating a series of explosions	• Assumption that metal will leak before thermal stress failure is not always valid. There were no temperature, pressure or gas detection upsets at BAHX prior to incident • No abnormal alarms during a shutdown prior to incident • Emergency shutdown and depressurization system manually activated 1 minute after initial explosion; however, damage permitted some feed to continue to flow and fire cause additional piping & equipment failures

The incidents summarized above share some common lessons that need to be considered when designing the SIS and SIFs:

- Process operations, such as filling, must have redundant systems that act to halt operations before loss of containment occurs. Process conditions can change rapidly, so field elements must have a rapid response time.
- Leaking LNG will create ice or hydrate as it vaporizes. Final element valves must be specified to function in the presence of these solids.
- On loss of containment, LNG will vaporize and has the potential to spread over a large area where it can encounter an ignition source. The SIS must rapidly detect the leak and perform system isolation to reduce the size of the release.
- Once loss of containment occurs, it is desirable to stop equipment, such as compressors, that adds energy to the system. This equipment must be stopped quickly and reliably while minimizing the potential for false trips.

3 SIS Design

LNG facilities are typically developed by connecting highly specialized process units often designed and constructed by multiple technology providers. The diverse technology packages, large and small, must be connected into one seamless operating facility that manages risk with a common philosophy. Vendor packages with specialized equipment, such as compressors, will

[2] Mixed Refrigerant
[3] Braised Aluminum Heat Exchanger

include a local Unit Control Panel (UCP) which serves as the package equipment BPCS to perform equipment operation, (i.e. surge control) and often there is also a SIS that performs equipment protection interlock (i.e. vibration shutdown). This will lead to the facility having a number of UCPs and unit SIS systems to manage in addition to the facility BPCS and SIS. Most project teams are highly focused on process integration, so the chemical composition, temperature and pressure interface points are well documented and managed, but primary control systems and package SIS may be viewed as stand-alone systems so formal integration is frequently not considered until commissioning and startup. Often package equipment providers design their process unit SIS consistently from project to project without considering the design of an adjacent package SIS or the facility SIS. The approach employed by each technology provider and equipment vendor will likely differ, creating an inconsistent risk management profile for the facility owner.

Establishing a consistent and effective SIS philosophy across the entire facility requires careful thought and clear specification to align all technology providers early in the design phase of a project. Developing an IEC 61511[8] compliant Functional Safety Management Plan (FSMP) is a critical step early in the project so that contracts can include the owner's interpretation of requirements for SIS design and implementation. Failure to provide this direction up front will permit vendors to execute according to their interpretation of the standard and will likely cause disconnects that become evident during commissioning or later. This can lead to costly redesign, start-up delays, and unmitigated risk exposure [9]. At a minimum, a project FSMP should establish the Functional Safety Philosophy sufficient to provide direction to vendors for the following concerns:

- Define safety and reliability requirements and preference for sensor and final element selection, including certification and reliability claim acceptance criteria.
- Certification requirements and useful life minimums for sensor and final element selection.
- Methodology and parameters to be considered in SIL Verification.
- Competency requirements for individuals performing SIS design and implementation tasks.
- Survivability requirements of a SIF in the event of a major accident event.
- Validation and system integration requirements.
- Proof test interval and testing philosophy (e.g. online vs offline).
- Reset philosophy.
- Safety Requirements Specification (SRS) philosophy.
- Facility wide Cybersecurity Risk Assessment and UCP/SIS requirements for cyber security concerns (i.e. threat barriers, update access).
- Functional Safety Assessment (FSA) requirements for all SIS systems.

The list above identifies FSMP content most pertinent to the early stages of the functional safety lifecycle leading up to start-up. The plan must also address operation, maintenance, modification, decommissioning, management and auditing concerns of the functional safety lifecycle.

4 Sensor and Final Element Selection

Equipment selection is one of the fundamental steps in the design phase of the SIS Safety Lifecycle. A SIF is composed of three main parts: sensing element, programmable logic solver, and final element. When selecting the devices for a SIF two sets of requirements must be considered for each component: functional requirements (suitability for application and operating environment) and safety integrity requirements (suitability for risk reduction).

Functional requirements relate to the suitability of the equipment to function properly in a given application. Some of these requirements can impact the selection of the programmable logic solver, but the functional requirements often have a larger impact on the selection of the sensing elements and final elements because they are exposed to the process environment. Factors such as SIF response time will impact the sensor, logic solver and final element selection[4].Typical questions that must be answered when addressing the functional requirements of a final element assembly are: is the valve the correct size, type, pressure class, material (i.e. severe service), and what is the actuation time? For sensors the conditions of the process connection (temperature, pressure, size, type, process interface material of construction) must be considered as well as the type of measurement, sensor location, type of signal to the BPCS, measurement accuracy/precision.

These questions for the sensing elements and final elements used in the LNG industry are particularly important due to the presence of cryogenic liquid, high pressures, and short response times. Different materials have different thermal sensitivities. In general ductility decreases as its temperature decreases, resulting in a higher probability of brittle fracture [10]. To protect against this failure mechanism cryogenic valves are typically made of specially selected metals and metal alloys to be resilient to low temperature and use extended stems and bonnets to improve performance in low temperature operations by increasing the volume of buffer gas.

Metallurgical concerns also apply to sensors in cryogenic LNG applications. Fortunately, many major equipment suppliers offer cryogenic options for common sensors allowing for easier specification of cryogenic capable sensors.

Most logic solvers and sensor elements have a sufficiently fast response for applications with a response time above one second; however, the response time for final elements, particularly considering the large line sizes and large inventories in LNG applications needs to be considered early in the design process. For example, although a response time of 10 seconds may seem long, when a valve is required to fully close in a 36" line may be difficult to achieve.

Safety integrity requirements are grouped into discrete Safety Integrity Levels (SIL) to identify the necessary risk reduction for the SIF. As the SIL Level increases from 1 to 4 the probability of failure of the SIF decreases. In the process industry typical applications, the highest required SIL level is SIL 3, with the majority of SIFs targeted at SIL 1 and SIL 2.

[4] Final elements are most often the limiting factor for achieving SIF response time.

Three numerical methods verifying that a SIF achieves the necessary SIL level are provided in Section 5 based on equipment failure rates and other factors.

In addition to selecting devices with adequate failure rates to achieve a given SIL level, the systematic capability must also be considered. Two methods are outlined in IEC 61511 for demonstrating systematic capability: Equipment Certification to IEC 61508[11] or prior use the justification.

If prior use justification is used, the amount of operating experience within a facility or organization for a specific component as well as performance tracking for that component is necessary. If the use of certified equipment is desired, this must be identified as a procurement requirement for components of the SIF. An example certification for a Cryogenic Final element is provided in Figure 1.

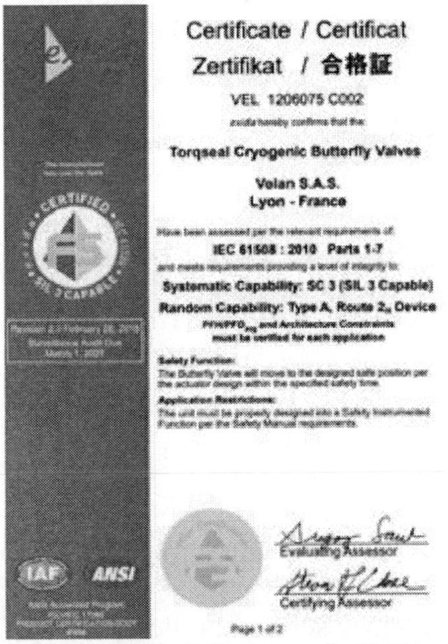
Certificate / Certificat
Zertifikat / 合格証
VEL 1206075 C002
Torqseal Cryogenic Butterfly Valves
Velan S.A.S.
Lyon - France
IEC 61508 : 2010 Parts 1-7
Systematic Capability: SC 3 (SIL 3 Capable)
Random Capability: Type A, Route 2H Device
must be verified for each application
Safety Function:
Application Restrictions:
Evaluating Assessor
Certifying Assessor
Page 1 of 2

Figure 1. SIL 3 Capable Certified Cryogenic Butterfly Valve [12]

When evaluating a certification for a device it is important to ensure that the certification is from an accredited body and that accurate information is provided for the equipment failure rate, systematic capability, and architecture requirements as each element will have an impact on the overall SIS design and verification activities.

Whichever method for meeting the systematic capability is chosen, due to the limited number of cryogenic safety-rated devices, special planning is required to ensure that both the functional and safety related requirements are met for all components of the SIF.

5 Verification

5.1 Verification Model

In section 3, SIL Verification methodology was identified as one of the many preferences that should be established in the FSMP. Calculating the average Probability of Failure on Demand (PFD_{AVG}) is one of the three hurdles that must be met to demonstrate SIL is achieved for SIF operating in low demand mode[5]. Probability of Failure per Hour (PFH) is equivalent metric for SIFs operating in high demand or continuous mode. The majority of SIFs in LNG applications are considered low demand mode. For the purpose of this paper, we will consider three generally accepted equation sets to calculate PFD_{AVG} for the example SIF. The primary difference between the equations is the number of variables considered.

The fundamental form is the most basic and is the foundation on which other forms are derived. This form considers only the test interval (TI) and the dangerous undetected failure rate (λ_{DU}) This equation assumes all other variables are "perfect" thus do not contribute to the dangerous failures.

$$PFD_{AVG} = \lambda_{DU} * \frac{TI}{2}$$

The simplified equation is a bit more complex and considers additional variables. This form additionally considers Proof Test Coverage (PTC), Mission Time (MT) and Mean Time to Restore (MTTR).

$$PFD_{AVG} = \left(PTC * \lambda_{DU} * \frac{TI}{2}\right) + \left((1 - PTC) * \lambda_{DU} * \frac{MT}{2}\right) + (\lambda_{DD} * MTTR)$$

A Markov model is actually a set of equations solved simultaneously considering the state of the system, considering degradation and repair, over a period of time. The theory of Markov model application for SIL verification is described in literature [13]. The Markov model for a 1oo1 system is visually depicted in Figure 2.

[5] A SIF is considered to operate in demand mode where the SIF is only performed on demand. SIF are considered low demand. If the demand is greater than twice the proof test interval, or more than once per year the SIF is considered to be operating in high demand and continuous mode criteria is to be applied.

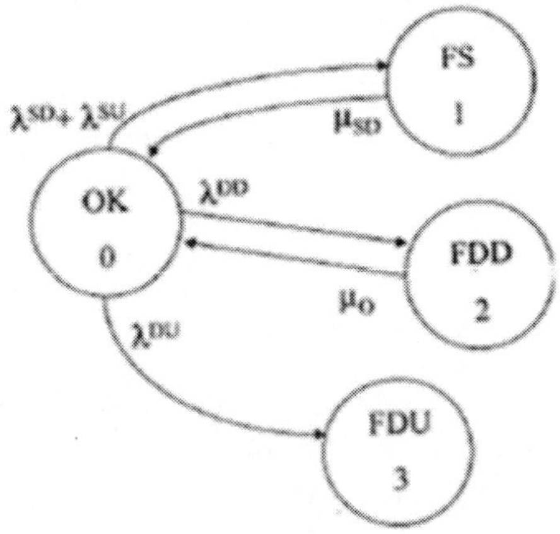

Figure 2. Markov Model for a 1oo1 System

For the purpose of this illustration, the calculations will consider variables which may significantly impact PFD_{avg} in a process industry application[14] such as LNG. Table 2 is a summary of the variables considered by the equations.

Table 2. PFD_{avg} Calculation Variables by Method

Variable	Fundamental Equation	Simplified Equation	Markov Model
Proof Test Interval (TI)	✓	✓	✓
Dangerous Undetected Failure Rate (λ_{DU})	✓	✓	✓
Dangerous Detected Failure Rate (λ_{DD})		✓	✓
Mission Time (MT)		✓	✓
Proof Test Effectiveness (PTC)		✓	✓
Mean Time to Restore (MTTR)			✓
Diagnostic Test Interval (DTI)			✓

Variable	Fundamental Equation	Simplified Equation	Markov Model
Proof Test Duration (PTD)			✓
Probability of Initial Failure (PIF)			✓
Maintenance Capability or Site Safety Index (SSI) [15]			✓

The end user must determine which equation best addresses the conditions of the application and recognizes variables that may be significant contributors to SIF failure. For example, as system where failed components are replaced may be well represented with the simplified equation, but in systems where components are repaired, and repair is imperfect may be better represented with by Markov model.

5.2 Verification Example

As demonstrated in the Esso Longford LNG terminal explosion loss of hot oil or other heating fluids can lead to dangerously low temperatures in critical process vessels. One possible SIF to address prevention or mitigation of this scenario is an independent low temperature interlock. For this SIF a low-temperature reading would initiate the closure of a cryogenic valve blocking process flow through the vessel to limit cooling below the minimum design temperature. PFD_{AVG} is calculated for an example SIF configuration using generic equipment failure rates for sensor and logic solver, and the cryogenic SIL-rated valve identified in Figure 1 as the final element. The failure rate and proof test coverage data for each SIF component is summarized in Table 3.

Table 3. Component Data

Device	**Notes**	λ_{SD}	λ_{SU}	λ_{DD}	λ_{DU}	**PTC**
Generic Temperature Sensor	Sensor	0	0	0	3E-07	90%
Generic SIL3 Certified PLC	Logic Solver	1.16E-05	1.29E-07	3.83E-06	2.41E-07	90%
Generic Solenoid Driver	Final element: Solenoid	0	3E-07	0	1E-07	73%

Device	Notes	λ_{SD}	λ_{SU}	λ_{DD}	λ_{DU}	PTC
Generic Rack & Pinion Actuator	Final element: Actuator	0	7.80E-07	0	4.20E-07	
Velan Torqseal Triple Offset Butterfly Valve	Final element: Valve	0	0	0	7.38E-07	
Total		1.16E-05	1.21E-06	3.83E-06	1.8E-06	80%

The example assumes that testing will be done offline, and that no diagnostics are used. Probability of initial failure is not considered. Other variables include:

- Mission Time (MT) – 30 years
- Mean Time to Restore (MTTR) – 24 hours
- Site Safety Index (SSI) – SSI 2: Good Repair

Table 4 compares results from the three equations.

Table 4. Results Comparison

Analysis	PFD_{AVG}	RRF	SIL (PFD_{AVG})
Fundamental Equation	7.88E-03	126.9	2
Simplified Equation	5.41E-02	18.5	1
Markov Model	6.44E-02	15.5	1

The SIL verification results for the example LNG application SIF demonstrate that the Fundamental Equation returns dangerously optimistic results primarily due to the unrealistic assumption of perfect proof test coverage. The calculated RRF in this method is higher by a factor of 8X than the Markov Model results and would lead to the incorrect conclusion that the SIF achieves SIL 2, when in reality it is only a relatively low integrity SIL 1 SIF.

The difference between the Simplified Equation and the Markov Models results for a simple SIF comprised of a single sensor element, final element, and logic solver is less significant, although the simplified equation is still optimistic. As the complexity of the SIF studied increases typically the potential discrepancy between the results of the Simplified Equation and the

Markov Models increases, making the use of a common SIL verification approach even more important to achieve consistent risk mitigation across the entire facility.

The Site Safety Index (SSI) is a measurement for assessing the overall maintenance capability of a given site, based on factors such as how frequently maintenance items are completed. The SSI and other considerations for the operations and maintenance phase such as the MTTR and duration of proof testing (if performed online) play a significant role in SIL verification calculations and are only incorporated as part of the Markov Model approach. These factors become increasingly more significant as a process becomes more integrated and proof test periods are extended.

6 Proof Test Planning & Execution

Periodic proof testing is a critical activity to assure that SIFs achieve the target reliability. The purpose of the proof test is to detect dangerous failures that any automatic diagnostics have not been able to find, so that repairs can be completed prior to a demand and confidence in the SIF performance restored. Requirements for IEC-61511 extend to the operations and maintenance, with responsibilities including: "routine proof testing of the SIF, calibration, repair, documentation, data analysis, and proactive replacement of components before end of useful life." [16] Successful fulfillment of these responsibilities requires adherence to the proof test schedule.

It is tempting to defer proof test planning until just before commission and startup; however, for SIS the proof testing decisions are fundamental to the design process. Proof test interval opportunity will likely be limited by turnaround frequency, unless SIFs are specifically designed to be tested on-line. The proof test method will directly impact the types and percentages of failures that can be detected by the proof test, thus setting the proof test coverage for components. As demonstrated above, both Proof Test Interval (PTI) and Proof Test Coverage (PTC) are significant variables in the verification calculations. Early consideration of proof testing methods and requirements as well as clear communication with operations and maintenance personnel will direct design decisions in order to provide the most flexible operating and maintenance opportunity.

Written proof test procedures are required to assure that the design commitments can be executed during the operation and maintenance phase of the safety lifecycle. These procedures, along with personnel training, will assure that the proof test is executed consistently and effectively over time, and that important failure information is collected for future analysis. The proof test procedure will embody the activities necessary to detect device failure and collect information that, when analyzed, confirms component failure rate and maintenance capability (site safety index). [17] As well as validate the SIF is capable of performing at the target SIL level. Data collected during the proof test must be retained and analyzed to confirm that the SIF reliability is within design parameters, or if modifications are required to meet SRS requirements.

7 Conclusion

This paper discussed some of the unique factors that impact SIS design in LNG processing facilities. We began by identifying the physical properties that contribute to the process hazards, and the project execution strategies that effect how the process hazards are managed. Several significant historical incidents were reviewed to illustrate some important industry lessons learned. The importance of a FSMP was illustrated and some key requirements for SIS design identified. Considerations for sensor and final element selection in potentially cryogenic service were provided. Results of common PFD_{avg} equations were compared to illustrate that the assumptions of a particular verification methodology must be considered to select the best method for a given process type. Finally, some operation and maintenance phase activities that are impacted by design phase decisions and provide confirming data to validate assumptions were discussed.

8 References

[1] Pipeline and Hazardous Materials Safety Administration (PHMSA), Available at https://www.phmsa.dot.gov/pipeline/liquified-natural-gas/lng-safety Accessed on February 19, 2020

[2] Property Hazards of LNG, Environmental Impact Statement, PNG LNG Project, https://pnglng.com/media/PNG-LNG-Media/Files/Environment/EIS/eis_attachment01.pdf Accessed on February 19, 2020

[3] *What Went Wrong? Case Histories of Process Plant Disasters*. Second Edition, Gulf Publishing, Houston, TX, 1988. pg. 113-115

[4] CCPS. *Incidents That Define Process Safety*. Center for Chemical Process Safety, American Institute of Chemical Engineers, New York, NY, 2008, pg 58-60.

[5] The Esso Longford Gs Plant Accident report of the Longford Royal Commission, Government Printer for the State of Victoria, June 1999. Available at https://www.parliament.vic.gov.au/papers/govpub/VPARL1998-99No61.pdf Accessed February 19, 2020

[6] CCPS. *Incidents That Define Process Safety*. Center for Chemical Process Safety, American Institute of Chemical Engineers, New York, NY, 2008, pg 120-121.

[7] U.S. Chemical Safety and Hazard Investigation Board, Loss of Containment, Fires and explosions at Enterprise Products Midstream Gas Plant, No. 2016-02-I-MS, February 13, 2019 Available at https://www.csb.gov/enterprise-pascagoula-gas-plant-explosion-and-fire-/ Accessed on January 19, 2020

[8] IEC 61511-1 Ed 2.0, "Functional Safety: Safety instrumented systems for the process industry sector – Part 1: Framework, definitions, system, hardware and application programming requirements," IEC, Geneva, Switzerland 2016.

[9] D. Chastain-Knight, R. Butz, and W. Donaldson. "Functional Safety Management Planning," Mary Kay O'Connor Process Safety Center, November 2017.

[10] C.J. O'Brien, L. Stewart, and L. Bredemeyer, *Final Elements in Safety Instrumented Systems*. exida, Sellersville, PA, 2018 pg 183.

[11] IEC 61508-1 to 7: 2010, "Functional Safety of electrical/electronic/programmable electronic safety related systems Parts 1-7," IEC, Geneva, Switzerland 2010

[12] exida Safety/Security Automation Equipment List (SAEL) available at www.exida.com/SAEL

[13] W.M. Goble and H. Cheddie, *Safety Instrumented Systems Verification, Practical Probabilistic Calculations*. ISA – Instrument Society of America, Research Triangle Park, NC, 2005 pg 275-301.

[14] I. van Beurden and W. H. Goble. "The key variables Needed for PFD_{avg} Calculations." Available at https://www.exida.com/Resources/Whitepapers

[15] Site Safety Index (SSI) Available at https://www.exida.com/SSI

[16] D. Chastain-Knight. "Functional Safety Practices for Operating & Maintenance," 14th Global Congress on Process Safety, April 22-25, 2018.

[17] D. Chastain-Knight and J. Jenkins. "Proof Test Prudently. Understand how to effectively evaluate low demand safety instrumented functions," Chemical Processing, September 2018, https://www.chemicalprocessing.com/articles/2018/proof-test-prudently/ (accessed February 19, 2020).

591523 Natural Gas and LNG Industry: Its Growth and Future Challenges

Abstract:

The past decade has seen rapid developments in multiple areas of the global natural gas value chain at a time of increasing concern for the environment. The low price of gas has benefited multiple industries including power generation and chemicals manufacturing, such as hydrogen production, as well as the transportation sector. Natural gas has significant environmental benefits over other fossil fuels, in particular coal. These benefits continue to drive the global demand for natural gas, and the development of new technologies. China's Blue Sky Initiative was responsible for the largest increment in global demand for LNG in the last few years.

Among the drivers for new technologies is the tightening of environmental regulations, such as the IMO (International Maritime Organization) 2020 regulations on sulfur content in marine fuels, which are typically heavy fuel oils. LNG represents an attractive replacement fuel and this law has generated a significant order book for LNG fueled new build ships, multi-role ships, and the development of specialized infrastructure for LNG bunkering. The demand for natural gas, both CNG and LNG, is also growing as a transportation fuel for trains and truck fleets to reduce sulfur dioxide, nitrous oxides, and particulate emissions. Conversion of natural gas to hydrogen will also likely play a role in the transition to a decarbonized future.

Growth in the use of natural gas also presents complex environmental challenges and opportunities - societal, regulatory, and technological. Natural gas is a fossil fuel as well as a potent Green House Gas. This has resulted in increased attention to flaring, venting, and fugitive emissions throughout the entire natural gas value chain followed by strides to develop innovative monitoring techniques. This presentation will review the developments in the value chain and discuss its current and future regulatory challenges.

591159 When Disasters Happen: A Collaborative Approach Towards the Community's Recovery Process

Abstract:

The 2020 Virtual AIChE Spring Meeting & 16th GCPS Wednesday keynote "When Disasters Happen: A Collaborative Approach Towards the Community's Recovery Process" will be presented by Thomas Munoz, Emergency Manager and Homeland Security Director with Texas City. As the Emergency Manager and Homeland Security Director with Texas City, the home of 19 chemical plants, he is responsible for the safety of all its citizens along with managing the crisis communication plan with local industry and major stakeholders.

Safety and disaster planning is vital to the recovery process after a major incident. As key critical infrastructure industries, issues may arise that have not been planned for in advance, such as the recovery from an incident. Developing and maintaining partnerships with local partners in your community will support addressing unforeseen challenges. Recovery efforts include planning with key partners and understanding the phases of recovery, which also consider the mental health and well of everyone involved in the response, as well as the community impacted. In this keynote address a discussion of pre-incident planning considerations and strategies as well as lessons learned/after action improvements post-recovery will be discussed. Additionally, collaboration considerations regarding local resources to promote recovery in the event of a workplace incident will be discussed.

C0-presenter Amanda Groller, MS.Ed., LPC will discuss the mental health impacts of a disaster.

591582 Plant of the Future

Abstract:

For the "Plant of the Future", there are a lot of promising options. Some of the future technology ranges from automated start-up, operations and shut-downs to Artificial Intelligence (AI) predicting process outputs and machine failures to remote locations being fully controlled due to internet connected sites. Some future changes could be process technological changes, such as electrification and closed loop economies. Another key point that will have to be addressed is ensuring site security and safety: both physical and electronic.

In this panel we will talk about opportunities and early successes and pitfalls and challenges. Topics will cover greenfield builds and brownfield expansions and retrofits.

INTEGRATED REFRIGERATION FOR GAS TURBINE DRIVEN LNG PLANTS

August 19, 2020

Xun Jin, Partha Ghosal, George Hu, and Rajeev Nanda
TechnipFMC

xun.jin@technipfmc.com

TechnipFMC

20AIChE Spring Meeting 16th Global Congress on Process Safety HOUSTON • 2020

Gas Turbine Inlet Air Chilling (IAC): Why?

- LNG Production ←→ Gas turbine driver performance
- IAC Benefits
  - Increase driver power/production
  - Improve GT fuel efficiency
  - Stabilize production
  - Flexible driver selection/configuration
- Require additional refrigeration

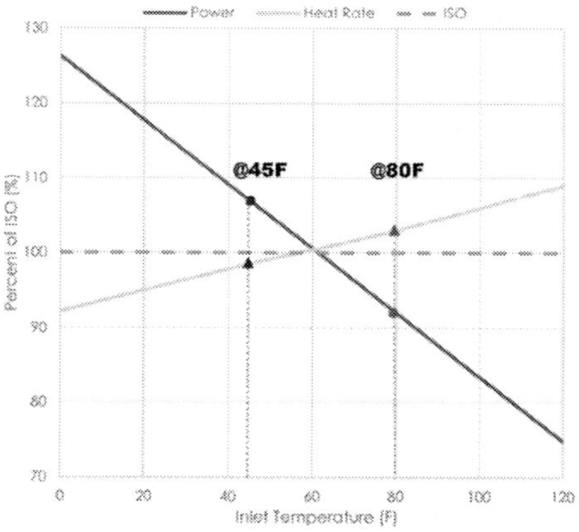

TechnipFMC

20AIChE Spring Meeting 16th Global Congress on Process Safety HOUSTON • 2020

Inlet Air Chilling: How?

- Mechanical or Absorption Refrigeration
- Current Mainstream Practice
  - Segregated mechanical refrigeration
- IAC LNG Application
  - One operating plant
- Why not Integrate with liquefaction refrigeration?

TechnipFMC

20AIChE Spring Meeting 16TH GLOBAL CONGRESS ON PROCESS SAFETY HOUSTON • 2020

IAC Integrated with Single Component Refrigeration

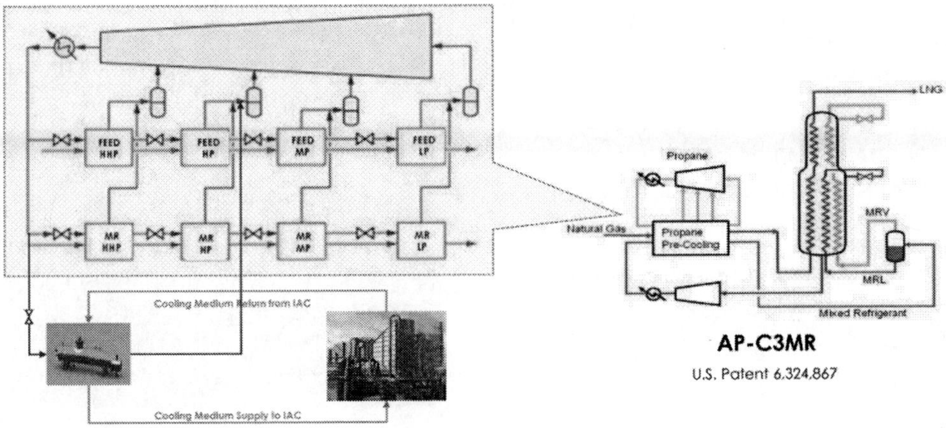

AP-C3MR

U.S. Patent 6,324,867

TechnipFMC

20AIChE Spring Meeting 16TH GLOBAL CONGRESS ON PROCESS SAFETY HOUSTON • 2020

IAC Integrated with Mixed Refrigeration

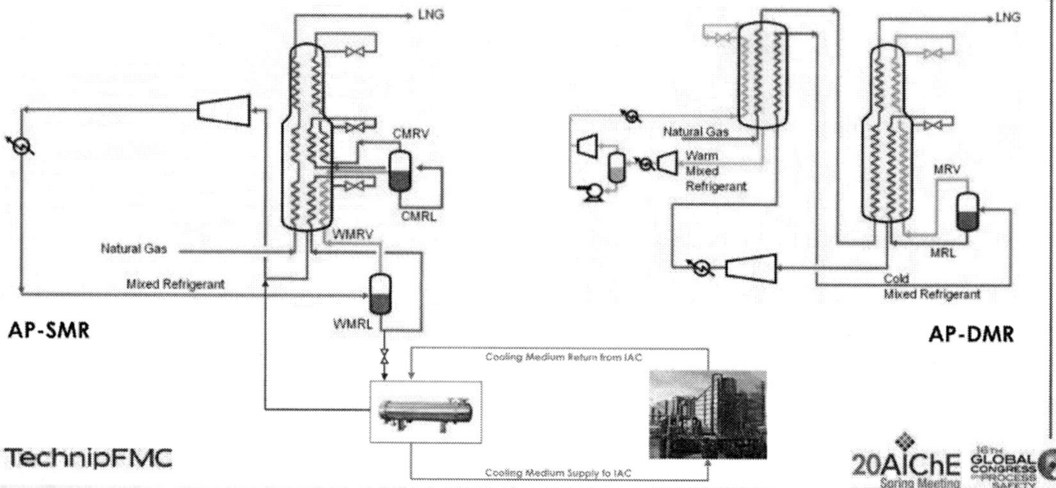

TechnipFMC

Comparison Case Study

- Segregated vs. Integrated Refrigeration for IAC
- Basis
  - 3 x 5MTPA C3MR trains
  - 4 x 54MW ISO Turbines per train
    - 2 for Propane
    - 2 for MR
  - IAC (Inlet Air at 46F) on MR turbines only
  - Ambient Air Dry Bulb: 89F
  - Ambient Air Wet Bulb: 81F

TechnipFMC

Segregated Design

- 6 x 3000 TR Propane Chiller Packages, each contains
  - 1 x screw compressor (3MW each)
  - 1 x propane receiver
  - 1 x Kettle propane vaporizer
- Air coolers
- A common glycol water loop
  - 2 x glycol water pumps

TechnipFMC

20AIChE Spring Meeting 16th Global Congress on Process Safety HOUSTON • 2020

Integrated Design - Propane Selection

- HP propane level was selected, i.e. vapor return to Compressor 3rd Stage Suction
  - LP Propane: - 33.9 F
  - MP Propane: -4.6 F
  - HP Propane: 23 F ⬅
  - HHP Propane: 58 F
- Propane draw location
  - Feed gas chiller train
  - Downstream of HHP chiller

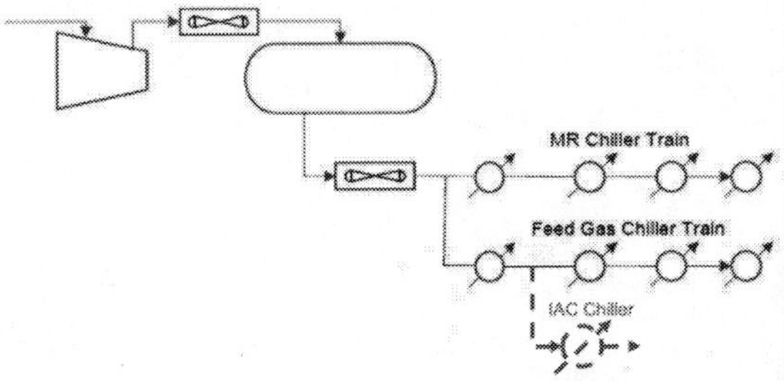

TechnipFMC

20AIChE Spring Meeting 16th Global Congress on Process Safety HOUSTON • 2020

Study Result Summary

- Based on this case study, Integrated Refrigeration:
  - Eliminated entire OSBL IAC refrigeration system
  - Saved **~$60 MM USD** in CAPEX
  - Reduced plot space
  - Reduced plant noise level
- **IAC Integrated with liquefaction offers a viable and potentially beneficial alternative to segregated refrigeration.**

TechnipFMC

20AIChE Spring Meeting 16th Global Congress on Process Safety HOUSTON • 2020

THANK YOU

xun.jin@technipfmc.com

TechnipFMC

20AIChE Spring Meeting 16th Global Congress on Process Safety HOUSTON • 2020

592114 Process and adsorbent for separating ethanol and associated oxygenates from a biofermentation system

Abstract:

Abstract for poster presentation or talk:

Abstract Title: Process and adsorbent for separating ethanol and associated oxygenates from a biofermentation system

Presenter Information: Deepak Sharma; Sr. Process Engineer, Engineering and Technology, Bayer AG

Patent number: US8658845B2

Patent Authors:

Deepak Sharma - Sr. Process Engineer, Bayer Ag, St. Louis; (Presenting Author)

Anil Oroskar – CTO, Orochem Technologies, Napervielle, IL;

David W. House – Scientist, Orochem Technologies, Napervielle, IL;

Alice M. Havill – Scientist, Lanzatech, New Zealand

The principles of process intensification if used on existing processes can lead to development and redesign of existing processes to become more sustainable. This presentation will discuss a process and an adsorbent for the separation of ethanol associated oxygenates from a dilute mixture of ethanol and associated oxygenates in water in the presence of organic compounds derived from a biofermentation process. After pretreatment, the separation is carried out in a simulated moving bed adsorption system employing a stationary phase adsorbent comprising fluorinated carbon or modified C18 silica gel selective for the adsorption of ethanol and associated oxygenates, such as 2,3-butanediol, with a mobile phase desorbent selected from the group consisting of methanol, ethanol, propanol, and methyl tertiary butyl ether. The process is useful for removing water from dilute aqueous mixtures of organic compounds comprising ethanol in dilute concentration in water and produced by fermentation, biomass extraction, biocatalytic, and enzymatic processes which are not economically recoverable by conventional distillation methods.

Keywords: Separation; Adsorption; SMB; Purification

Patent Assignee: Orochem Technologies LLC

592111 Pump safety, the undiscussed topic in chemical industry leading to several fatal incidents

Abstract:

Abstract Title: Pump safety, the undiscussed topic in chemical industry leading to several fatal incidents

Presenter Information: Deepak Sharma; Sr. Process Engineer, Engineering and Technology, Bayer AG

Pumps are one of the equipment's in the process industry that require a thorough review during HAZOP studies. The constant source of energy given to the pump, if not converted to kinetic energy to flow the fluid, can lead to heating of the pumped fluid. If the fluid keeps accumulating the energy provided by the motor, in a blocked-in scenario, it can lead to very high temperatures eventually leading to loss of containment of the pump and the pump can get damaged. There have been instances in the industry where fragments of damaged pumps have flown several feet away from the point of loss of containment leading to injuries. The poster will outline a scenario where a pump is run blocked in on suction and discharge for water and a chemical which has a low thermal decomposition temperature. Arc data is used to evaluate the onset temperature of thermal decomposition. The poster will outline safeguards that can be employed to avoid catastrophic loss of containment of the pump. Outlined below are safeguards used to avoid loss of containment of a pump in case of a blocked-in scenario:

- Install PSV on suction of pump
- Employ a low amp/power interlock on the pump to turn off the pump in case of blocked in situation
- Have a high-pressure interlock on the discharge line of the pump to direct flow back to the tank in case of pump deadhead
- Install an Interlock to turn off the pump in case the feed forward valves (automatic valves) close inadvertently

Pump deadhead scenarios are not only catastrophic when in flammable service there are several salts that have low thermal decomposition temperatures. Pumps in water service left blocked in could also blow catastrophically leading to reportable incidents. I feel this is an important topic to bring to the attention of students and professionals in the chemical industry. Poster will include pictures of actual installed pumps along with required safeguards in the field along with supporting P&ID depictions.

589440 New Kind of Linear Nanocomposite Applied in Improving Rheological Behavior of Crude Oil and Its Mechanism Studied through Molecular Simulation Method

Abstract:

In this work, a new kind of linear nanocomposite produced by carbon nanotube(CNT) as pour point depressant (C2-100) was developed by blending with MCNT nanoparticle and copolymer named Ethylene-vinyl acetate (EVA). Its effect of reducing yield stress and modifying the morphology of wax agglomerate were also evaluated. Besides, the synergetic effect of C2-100 nanocomposite with the magnetic field was also conducted. The mechanism of modifying rheological behavior of waxy crude oil was studied through molecular simulation method, which aims at understanding the interaction between this nanocomposite and wax molecule. The results indicated that C2-100 nanocomposite could reduce the yield stress of waxy crude oil more effectively than the copolymer (EVA) even at low concentration, as well as changing the morphology of wax crystals and agglomerate. Moreover, its effect of reducing yield stress becomes more obvious when the alternate magnetic field was applied. In addition, the results of molecular simulation also presented that the effect of improving rheological behavior of C2-100 nanocomposite is related to the adsorption behavior of wax molecules on the surface of MCNT nanoparticle, which is adsorbed with EVA molecules. In addition, the adsorption behavior of different fragments of EVA molecules can also influence the adsorption behavior of wax molecules on the surface of EVA-attached MCNT nanoparticle, which may influence the effect of C2-100 to improve the rheological behavior of waxy crude oil.

589683 Solar Energy Conversion and Noise Characterization Sensing in Building Integrated Solar Photovoltaic Devices with Ventilation

Abstract:

A study on solar energy conversion and noise characterization sensing in solar photovoltaic devices with ventilation is conducted. A building integrated parallel plate solar photovoltaic (PV) device was built with a pair of PV modules, a ventilated air cavity and an insulating back panel of plywood board filled with polystyrene installed in an outdoor test-room. Photovoltaic amplification was attained with power output from a potentiometer through rotation of its circular knob. A parallel plate PV device was studied for its
electrical parameters as Resistance-Capacitance (RC) electrical analog circuit. The effect of inductive and capacitive heating losses was considered in evaluating electrical characteristics of a PV device exposed to solar radiation. The characterization of noise interference, due to power difference of two intensities in a wave on PV Device is presented. Noise characterization of a wave is obtained depending on type of wave. The difference of two power intensities is due to transmission of light, sound, heat, electricity, fluid, and fire into a parallel plate PV device. Based on analysis of a solar PV device, a definition with measurement expression and unit for an electrical noise were coined. Standard definitions of noise sources, their measurement equations, their units and their origins under limiting reference conditions were also devised. A review for sensing noise of physical agents is conducted for sensing noise of physical agents (sun, light, sound, heat, fluid, electricity and fire). With integrated environmental control, these environmental parameters are to be characterized through various environmental monitoring sensors.

589945 Structural Characterization and Model Representation of Vitrinite-Rich India Coal By Using Advanced Analytical Techniques

Abstract:

The detailed investigation of the chemical structure of Karharbari coal (vitritine rich subbituminous coal from India) by utilizing advanced instrumental analytical techniques such as solid state ^{13}C nuclear magnetic resonance (^{13}C-NMR) spectroscopy, Fourier transform infrared (FT-IR) spectroscopy, X-ray diffraction (XRD), Flash-pyrolysis gas chromatography mass spectroscopy (Py-GC/MS), Laser desorption ionization time-of-flight mass spectroscopy (LD-TOF-MS), X-ray photoelectron spectroscopy (XPS), and high-resolution transmission electron microscopy (HRTEM). The Karharbari coal was an amorphous loosely organized lower rank coal with aromaticity 0.79. This was indicated by XRD results and also confirmed with ^{13}C-NMR results (aromaticity =0.84) and FTIR results (aromaticity = 0.82). The average molecular weight of the aromatic cluster is 507 amu by NMR result which is closer to the result obtained by HRTEM (MW = 530.46 amu). The presence of different heteroatoms like carbon, oxygen, nitrogen and sulphur with their functionalities, were determined by using a XPS technique. Different carbon/oxygen forms present in the Karharbari coal are (C-C) and (C-H) 68.53%, (C-O) 23.46%, (COO^-) 1.96%, (C=O) 6.05% and different nitrogen forms such as pyridine, pyrrolic, quaternary, oxidized and their distribution (mol %) are 19.36%, 45.68%, 31.25%, 3.71 respectively, and different sulphur forms were present i.e. corresponding to thiophenes, sulphones, sulphuric acid and sulphates with the molar content of 16.47%, 41.61%, 21.37%, 20.55% respectively. The construction of the individual molecules will be discussed in detail in the rest of this paper. The individual molecules were geometry-optimized and Materials Studio Amorphous Builder module was used to construct a three dimensional associated structure.

589479 Life Management of High Energy Piping with Highlights from ASME B31.1

Abstract:

High energy piping (HEP) systems – or as defined by the ASME B31.1 Power Piping Code, Covered Piping Systems (CPS) – are critical assets with respect to production facilities. A failure can have catastrophic consequences from a safety perspective and in terms of equipment loss and downtime. In fact, loss of high-pressure steam due to a failure in a main run of steam supply piping could shut an entire production unit down. Because of this and the requirements of the Code, HEP maintenance and inspection programs are established to ensure that the integrity of the systems is maintained in a safe and reliable condition throughout their lifecycle. There is no one-size-fits-all for the life management of these critical assets because there is a great variety in age, design, materials, and owner objectives. Life management of these assets is beneficial in that it provides a structured, systematic approach to creating and implementing a program including evaluating the current program (whether formal or informal), what the objectives of the program are (short and long term), what equipment is in place, and its history.

A HEP program put in place for overall life management is not synonymous with an inspection program, although inspections are an important part of an overall program. Inspections should be used in conjunction with analytical tools, real time monitoring, and laboratory examinations. The overall approach is life management – not inspection management – of the equipment.

This paper describes a HEP life management approach which includes identifying locations of interest, risk-ranking those locations, implementing a plan for determining the health of the highest risk locations, and taking actions based on the system health that has been determined. This paper also presents and discusses some of the requirements of the B31.1 code, which at the highest level states that each operating company shall develop operation and maintenance procedures for piping systems deemed necessary to ensure safe facility operations based on the provisions of the code, relevant industry experience, the operating company's experience (and knowledge of its facility), and the conditions under which the piping systems are operated.

590131 Optimization of C3MR LNG Liquefaction Process Using a Parallelized Simulator

Abstract:

Controlling the cost of energy-intensive processes such as C3MR (propane/mixed refrigerant) LNG liquefaction is a prime opportunity for optimization. Our work combines different C3MR models to develop an optimization that uses multiple independent variables, relaxed constraints, and flowsheet alterations to achieve an optimal operating cost. This work uses a rigorous model of the process in a process simulator rather than a reduced model of the process response. Use of a first-principles process simulator increases the accuracy of the model and therefore allows a more realistic optimization, at the cost of increased computing time. Parallelization (using multiple CPU cores) in the process simulator greatly increases the speed of the optimization calculations. CHEMCAD's optimization tools and multi-core technology make it possible to easily define the optimizations, run the optimizations efficiently, and generate global optima with a high degree of confidence.

590196 Modifying Fischer-Tropsch Reaction Sequence with Addition of Zeolite to Iron-Based Activated-Carbon Supported Catalyst

Abstract:

In addition to previously presented scientific work, modified reaction sequence of Fischer-Tropsch (FT) process was studied by addition of Zeolite ZSM-5 to an iron-based activated-carbon-supported catalyst. Main reactions involved due to FT catalyst are Fischer Tropsch Synthesis (FTS) and Water Gas Shift (WGS) reactions. Main reactions due to ZSM-5 are cracking, oligomerization, isomerization, cyclization, aromatization, conversion of alcohols to hydrocarbons, and coke formation. The waxy and linear structure FT products are converted to gasoline range, lighter hydrocarbons, iso-paraffins, and aromatics using ZSM-5. The catalyst ratio of 1:2, FT catalyst: ZSM-5, is used at 300 PSIG and 300°F. Separate-bed configuration, i.e. FT and ZSM-5 are separated in a reactor, arrangement is used for the reaction sequence analysis. Alcohols with carbon-chain from 3-5 produced by FTS are converted into hydrocarbons over ZSM-5. Waxy hydrocarbons with carbon-chain greater than 23 produced by FTS are cracked and isomerized to small n-paraffins and iso-paraffins using ZSM-5. The hydrocarbon with carbon chain from 7-10 produced by ZSM-5 cracking reactions are converted to aromatic products over ZSM-5. Produced gasoline to kerosene range n-paraffins are further cracked into smaller chain (1-5) iso- and n-paraffins. Some of the iso-paraffins are oligomerized and aromatized over ZSM-5 to form more aromatics. Further, multiple aromatic rings combine to undergo cyclization and dehydrogenation reactions over ZSM-5 to form coke. Water produced as a byproduct of FTS is combined with some of the carbon monoxide, which is the WGS, to form carbon dioxide and hydrogen.

590427 Lateral Retention Force Required for Sliding Saline Droplets over Silianized Silicon Surface

Abstract:

The lateral retention force required to slide a liquid drop on a surface is expected to increase with the time that the drop rests on the surface prior to sliding till it reaches a plateau. The molecular reorientation of the molecules at the periphery of the solid surface subjected to the unresolved vertical component of surface tension force causes this phenomenon. In this system, the surface tension of deionized water is increased by adding NaCl and is expected that the lateral retention force is increased proportionally with concentration. The retention force increased with increase in concentration reaching a maximum only to drop to a minimum with concentrations closer to saturation. That is, the lowest sliding force is required for both lower and higher concentrations of NaCl. This sudden decrease in observed lateral force is not a function of surface tension and the surface on which the drop slides. This phenomenon is associated with adhesion of salt to the surface and is supported by the drop sliding motion and depinning of the liquid from the solid.

590433 Study of Droplet Motion across Polymeric Substrates By Centrifugal Adhesion Balance (CAB)

Abstract:

On a given surface, the contour of the drop changes accordingly based on factors of the force applied, chemical structure of the fluid, drop size, and surface topography. With increasing force that is exerted on the drop, we determine the comprehensive trend for drop motion with respect to the controlled factors mentioned above. Drop motion is categorized into four phases and the point at where the initial displacement of receding line is observed is defined as the onset of motion. In this study, the onset of motion is affected by the factors mentioned above and aids in the determination for another value correlation known as Gs, known as Interfacial Modulus. The resistance to interaction towards the droplet as observed is quantified with respect to change in surface topography. This study therefore characterized drop motion and specifically onset motion based on controlled variables (factors). Work of adhesion is also quantified across different substrates to characterize the detachment of liquid from its substrate. All these parameters such as Gs, onset of motion, contact angles were analyzed with different polymeric substrates, drop sizes, and various fluids used.

590665 Population Balance Modeling and Spatially-Oriented Control for Crystal Size and Shape in Crystallization Processes

Abstract:

Crystallization is an important separation and purification process with relevant applications in the food, pharmaceutical, fine chemical and fertilizing industries. In these processes, modeling and process control for crystallization are focused on crystal size distribution (CSD) and shape of the crystals in order to make it possible to achieve the desired quality and purity of the product of interest. The control of crystal size and shape is of considerable importance for industrial crystallization processes, as crystal morphology has great influence on downstream steps and also from the point of view of product specification criteria. The consideration of crystal shape in modeling of crystallization processes is also important due to its impact on product effectiveness, such as the bioavailability and tablet stability in pharmaceuticals. Control of the crystallization process is can be facilitated with online measurements, which provide information to be used in the modeling of the process. However, if online measurements are not available, the strongly nonlinear dynamics of these processes makes difficult to establish relations between some inputs (e.g. temperature) and the final desirable characteristics of the crystal product. In this context, image analysis has been shown to be a suitable technique to monitoring the process [1].

In our present work, a population balance model-based scheme for controlling the size and shape of crystals is presented. The modeling was performed using a two-dimensional population balance (2D-PBM) in terms of the two characteristic dimensions of the crystals, describing the evolution of crystal size and shape distributions along the batch time. To identify the nucleation, growth and dissolution kinetics parameters using the 2D-PBM approach along the length and width directions, QICPIC-LIXELL's high speed dynamic image analysis equipment was used for the online measurement of the CSD, shape parameters and aspect ratio distribution of crystals during the experiments. In this way, the developed model allowed for the prediction of the crystal size and shape distributions during growth-dissolution cycles.

The developed framework was employed to obtain optimal temperature policies for controlling the crystal size and shape distributions that meet target criteria for size, morphology and productivity using the dynamic image analysis along the runs. Based on the concept of using measurements along batch crystallization processes as a trajectory in a phase space [2], a trajectory-endpoint control problem is presented with an oriented control scheme that is applied to the shape of the crystals. The scheme was used to control the batch cooling crystallization of potassium dihydrogen phosphate. The reported experimental data demonstrate the application of the control policies for temperature to produce crystals of the desired average shape and size in different batch times. Thus, using the first principle modeling coupled to the developed trajectory tracking control, the proposed framework has the potential to be applied to industrial crystallization processes for shape control and to be extended to polymorph control investigations.

References:

[1] Eisenschmidt, H.; Bajcinca, N.; Sundmacher, K. (2016) Optimal Control of Crystal Shapes in Batch Crystallization Experiments by Growth-Dissolution Cycles. *Crystal Growth & Design*, v. 16, p. 3297-2803.

[2] Griffin, D. J., Grover, M. A., Kawajiri, Y., Rousseau, R. W. (2015) Mass-count plots for crystal size control. *Chemical Engineering Science*, v. 137, p. 338-351.

590450 Auditing the Process Hazard Analysis - Key Characteristics to Scrutinize in a Good-Quality HAZOP Study and Supporting Effective HAZOP Study Updates

Abstract:

Although Process Safety Management (PSM), requiring good-quality Process Hazard Analyses, was promulgated in the United States in 1992, and HAZOP Study techniques date back to the mid-1970's, the HAZOP Study technique is still not uniformly applied in a high quality fashion. The objective of this paper is to provide experience-based tips on Process Hazard Analysis auditing, and some key characteristics that the auditor can scrutinize to assure quality implementation. Analogously, a similar set of characteristics that can facilitate future HAZOP Study updates can also contribute to a quality HAZOP Study that supports the spirit of Safety Management System objectives for the identification of hazards potentially impacting plant personnel, the community, and the environment. This paper will also review tips for structuring the HAZOP Study to facilitate later updates and revalidation.

590342 Development of High Performance Organic-Based Cathodes

Abstract:

For lithium-based batteries, structure/property relations in terms of cell voltage are desired for new organic-based cathode active materials. In this initial pursuit of structure/property relations, ab-initio calculations were performed on a few organic-based molecules to show the effect on the theoretical cell voltage by inclusion of nitrogen and/or oxygen. Preliminary test results are presented on a newly developed molecule and methods to potentially improve on the cell performance are given.

590705 Electrochemical Kinetics of Degradation of Graphite Anodes in Lithium Ion Batteries

Abstract:

Growth of the solid electrolyte interphase (SEI) is a major driver of capacity fade in LIBs. Despite its importance, the fundamental mechanisms remain unclear, primarily because of the complicated reaction pathways involved[1–3]. SEI growth can be both electrochemical and chemical in nature[4], and thus, it is a strong function of the potential and degree of lithiation of the electrode. We model the early-stage and long-term growth of SEI by accurately capturing the potential dependence of its formation kinetics as well as long term rate limiting steps, and validating it against the world's largest open source battery cycling data, generated in-house[5]. This is done using the Multiphase Porous Electrode Theory (MPET) framework[6] on graphite (phase separating) and carbon black (non phase separating) particles.

Lithium plating is another key degradation phenomenon that has been elusive, and it becomes important while trying to fast-charge batteries, i.e., 0% - 80% state-of-charge in 30 mins. We show that lithium plating is a key function of electrode morphology, phase-separation dynamics and potential. Phase-separation in graphite is modeled in the electrode using the Cahn-Hilliard Reaction framework described by Bazant[7]. We understand the electrochemistry of the onset of lithium plating with in-situ measurements connected to real time cell potential in a phase-separating electrode for the first time[8].

Results indicate that the peak SEI-forming currents are higher for higher driving currents. Also, we find that SEI only grows during electrode lithiation, i.e. the battery only degrades while being charged. We also find that onset of lithium plating is correctly captured only when phase separation in active material is accounted for. Further, the onset of plating is delayed on electrodes with a thick SEI layer – understanding SEI/plating coupling is integral to predicting fast charging manufacturing protocols for LIBs. This work holds promise for the predictive design of procedures[9] for manufacture and formation of LIBs.

[1] Cohen, Y. S.; Cohen, Y.; Aurbach, D. Micromorphological Studies of Lithium Electrodes in Alkyl Carbonate Solutions Using in Situ Atomic Force Microscopy. J. Phys. Chem. B 2000, 104, 12282–12291. https://doi.org/10.1021/jp002526b.

[2] Horstmann, B.; Single, F.; Latz, A. Review on Multi-Scale Models of Solid-Electrolyte Interphase Formation, Current Opinion in Electrochemistry 13, 62-69 2019.

[3] Nie, M.; Abraham, D. P.; Seo, D. M.; Chen, Y.; Bose, A.; Lucht, B. L. Role of Solution Structure in Solid Electrolyte Interphase Formation on Graphite with LiPF6 in Propylene Carbonate. J. Phys. Chem. C 2013, 117 (48), 25381–25389, https://doi.org/10.1021/jp409765w.

[4] Das, S.; Attia, P. M.; Chueh, W. C.; Bazant, M. Z. Electrochemical Kinetics of SEI Growth on Carbon Black: Part II. Modeling. J. Electrochem. Soc. 2019, 166 (4), E107–E118. https://doi.org/10.1149/2.0241904jes.

[5] Attia, P. M.; Das, S.; Harris, S. J.; Bazant, M. Z.; Chueh, W. C. Electrochemical Kinetics of SEI Growth on Carbon Black: Part I. Experiments. J. Electrochem. Soc. 2019, 166 (4), E97–E106. https://doi.org/10.1149/2.0231904jes.

[6] Smith, R. B.; Bazant, M. Z. Multiphase Porous Electrode Theory, J. Electrochem. Soc. 2017, 164 (11). https://doi.org/10.1149/2.0171711jes.

[7] Bazant, M. Z., Theory of chemical kinetics and charge transfer based on nonequilibrium thermodynamics, *Accounts of Chemical Research*, *46*(5), 1144–1160. https://doi.org/10.1021/ar300145c

[8] T. Gao, Y. Han, S. Das, T. Zhou, D. Fraggedakis, N. Nadkarni, C. N. Yeh, W. Chueh, J. Li, M.Z. Bazant, Interplay of lithium intercalation and plating on graphite using in-situ optical measurements, submitted.

[9] Huang, W.; Attia, P. M.; Wang, H.; Renfrew, S. E.; Jin, N.; Das, S.; Zhang, Z.; Boyle, D. T.; Li, Y.; Bazant, M. Z.; McCloskey, B. D.; Chueh, W. C. and Cui, Y.; Nano Letters 2019 19 (8), 5140-5148. DOI: 10.1021/acs.nanolett.9b01515

589787 Model Development of Solid Oxide Fuel Cell Thermal Performance Using Artificial Neural Network

Abstract:

NASA's Johnson Space Center has recently begun efforts to eventually integrate air-independent Solid Oxide Fuel Cell systems, with landers that can be propelled by LOX-CH_4, for long duration missions. Using landers that utilize such propellants, provides the opportunity to use such systems as a power option, especially since they will be able to process methane into a reactant through fuel reformation. To ensure fuel reformation in the systems, Steam Methane Reformation (SMR) will be employed. Various lead-up activities, such as hardware testing and computational modelling, have been initiated to assist with this developmental effort including the development of dynamic empirical models using artificial neural network. Such a modeling approach has shown to be accurate (R^2>0.9) to predict the thermal and electrical performance for SOFC systems. Hence, such models could be used for system optimization and control design.

Impact of Spacer Nature on Rheological Properties of Cationic Gemini Surfactant

Abstract ID: 589650

Shams Kalam[1], Muhammad Shahzad Kamal[2], Shirish Patil[1] and S. M. Shakil Hussain[2]

(1) KFUPM, Dhahran, Saudi Arabia

(2) Center for Integrative Petroleum Research, King Fahd University of Petroleum & Minerals, Dhahran, Saudi Arabia

Surfactants play a vital role in Chemical Enhanced Oil Recovery (cEOR) techniques. Gemini surfactants are a unique style of surfactants having more water-soluble hydrophilic head and oil-soluble hydrophobic tail units. Both groups are connected by a spacer at or near to the head groups. Polymers are added into the surfactants to increase the sweep efficiency of the flooding processes. This paper presents the rheological properties of Surfactant Polymer (SP) solutions in which two recently synthesized cationic gemini surfactants with different spacer nature are used.

The rheological properties of novel SP solutions were determined using Discovery hybrid rheometer (DHR-3) from TA Instrument. The geometric configuration set for this work was concentric cylinder. All experiments were carried out at 80 °C, while the shear rate for steady shear viscosity measurements ranges from 0.001 to 1000 s^{-1}. 0.25 weight % of a polymer FO 4290 SSH was used with different concentrations of newly synthesized gemini surfactants for Rheological measurements. Two hours magnetic stirring and twenty hours retention time were provided to make uniform surfactant polymer solutions.

Both surfactants exhibited good thermal stability in seawater with no precipitation. Shear viscosity and storage modulus were measured as a function of shear rate and angular frequency respectively at different surfactant concentrations ranging from 0 to 0.5 mM, where 0 mM shows pure polymer. We discovered that counter ions also influence the rheological properties significantly. Research outcomes revealed that the novel SP solution with a benzene ring and chlorine counterion produces better performance in comparison with benzene ring and bromine counterion. Moreover, the effect enhances if we increase one more benzene ring in the spacer.

This study reports two recently synthesized SP solutions for cEOR processes in carbonate reservoirs. It presents the effect of counterion on rheological properties and improves the understanding of tuning the surfactant structure in making SP solutions to increase the flooding sweep efficiency.

590257 Real Time Energy Management Systems Implementation for Continually Optimized Performance

Abstract:

It is well known that energy management is key for the optimal operation of refining and petrochemical sites. Industrial energy systems are constituted by complex utilities networks that are interconnected and have complex interactions that in order to achieve it best performance need to comply with increasing environmental rules, push against operating constraints and adapt to altering power prices and rapidly changing equipment availability.

After the initial installation, management and continuous improvement of these systems present challenges not only from a technological but also from an organizational point of view.

The implementation of an advanced digital Energy Management System (EMS) which includes capabilities for multi-period optimized scheduling, real time optimization and monitoring, deployed either as cloud-based or on-premise solutions that could be acting on open or closed loop is known to provides large benefits.

Critical is the sustainability of the EMS in the ever-evolving energy supply and demand of refining and petrochemical sites. Commercial, asset and organizational factors all change and the EMS needs to adapt to the developing environment of each enterprise meaning the type of implementation, including appropriate management structure, culture, best practices and Key Performance Indicators (KPIs) needs to be continually improved.

The EMS implementation at the large and complex Eastman site in Longview, TX, will be presented as an example. It will provide insight on the initial successes achieved in real time energy management for both, cost optimization and energy related KPIs performance tracking. The case study evolves to discuss how technical asset improvements, once replicated in the EMS, provided alternative operating modes, large additional benefits and sustained optimal

Andrew Lammers, NN2

Eastman Longview

Pablo Montagna*, Cristian Rojas

KBC Advanced Technologies, Inc

589818 Analysis of Fuel Cell System through Project Based Learning for Application of Classical Thermal Fluid Concepts in Mechanical Engineering Laboratory Course

Abstract:

A hands-on project involving analysis of air noncircular channels in Fuel Cell to measure the pressure was introduced in a thermal fluid laboratory course. Although the students studied fluid mechanics as a theoretical course, this project gave them the opportunity to build an experimental setup from scratch to help showcase their hands-on skills and to apply their understanding of relevant concepts of fluid sciences. The results of the project included a successful test of a conceptual procedure, designed to measure and calculate pressure drop and flow rates across the anode and cathode sides of a fuel cell. The students enjoyed completing and learning from the project. They also provided very useful feedback on the project by incorporating a functional experimental system for the future laboratory project. This project based learning opportunity allowed students to build their confidence in what they are learning as well as provide a glimpse of what the real world may hold.

589438 Measured Versus Calculated Sour Gas Properties

Abstract:

Measured versus calculated sour gas properties
Abstract:
Recent Environmental agencies have restricted standards for sulfur emission from gas chemical processing. Sour gas can extremely impact the delivery and processing by corroding the equipment. Gas properties such as compressibility, density and viscosity are a major factor controlling prediction of future production pipelines, and production tubing and pipelines in gas industry.
Sour gas effect on all gas industry aspects usually gas properties are measured in laboratories but occasionally experimental values are not available due to complexity and time consuming hence a reliable method would be proposed.
We are assessing the accuracy of the industry standard methods in calculating sour gas properties, compressibility factor using the following correlations:
Drunchuck - Abukkasem, Hall – Yarbarough, Drunchuck- Purvis- Robinson, Londono et-al, Alanazi et-al and Azizi et-al.
Viscosity using empirical correlations such as Carr-Kobayashi Burrows, Lea-Gonzalez –Eakin and Dean- Stiel.
Density using equations of state Redlich Kwang and Soave -Redlick - Kwang.
This study shows that the best correlation for calculating compressibility factor and density of high concentration sour gas is Londono-et al, all other considered methods are not accurate. All other methods investigated include high error especially for high H2S concentration.
However the error in estimation of Z factor for most of the considered methods has great impact on estimated sour gas density. Therefore we recommend that sour gas properties be measured experimentally at high pressure and high temperature conditions.

587967 The Study of a S-CO2 Oxy-Fuel Combustion Power Plant for Low Grade Coals

Abstract:

Recently, it is very important to have reliable solutions that can maximize power generation efficiency of current power generation processes and decrease emissions of CO2 and other pollutants because there are increasing concerns about the global warming and the depletion of fossil fuels. Among the available carbon capture technologies and power generation technologies, a S-CO2 (supercritical carbon dioxide) cycle based oxy-fuel combustion process appears to be a promising mid-term solution for the production of electricity from fossil fuels. Since supercritical carbon dioxide (S-CO2) involves a rapid density change at small temperature and pressure ranges, the S-CO2 cycle can extract large amounts of energy from relatively small equipment and it is possible to minimize or simplify a power generation block and to obtain higher thermal efficiency than existing steam cycles used in power generation plants. Therefore, an indirect S-CO2 oxy-fuel combustion power plants for low grade coals was designed and analyzed by using ASPEN-PLUS. Also, the operating characteristics of the indirect S-CO2 Oxy-fuel combustion power plant were compared with those of conventional air combustion power plants.

590099 Characterization and Performance Assessment of Oil Entrapment By Commercially Available Surfactants

Abstract:

Oil spills occur frequently and are a threat to the health and safety of human society and the well-being of the environment and ecosystem. Remediating oil spills is an ongoing challenge. Surfactant based dispersing agents, such as the Corexit® product line, are widely used for remediation of oil spills mainly through direct application on top of the oil slick. This work encompasses the investigation of variables that influence the emulsification process and therefore the encapsulation of oil within the surfactant micelles. Of central importance are the conditions under which oil entrapment in emulsions even occurs. A central finding is that there appears to be a threshold shear rate for emulsifying oil with surfactants, below which no emulsification occurs. Different schemes for applying shear stress in this work included simply stirring, using a high energy colloid mill, and flowing the mixture through a high shear capillary. Also, the sequence in which the oil/surfactant/water (O/S/W) components are mixed impinges strongly on whether and how much oil entrapment occurs. To try to correlate shear stress and velocity profiles throughout the sample volume with entrapment efficiency and kinetic stability of these O/S/W systems, Computational fluid dynamics (CFD) simulations were conducted. Another important factor is the ionic strength of the aqueous medium, which greatly impacts kinetics of these emulsions, but with a smaller impact on the threshold at which any emulsion is obtained at all. Surfactants tested include Corexit® EC9500A, sodium dodecyl sulfate (SDS), Triton X-100, and cetyltrimethyl ammonium bromide (CTAB). The quantification of oil entrapment was performed through either thermal gravimetric analysis (TGA) or differential scanning calorimetry (DSC).

Designing a Native Model for Assessment of the Effectiveness of Advertising

Sorush Niknamian
PhD, military Medicine Department, Liberty University, USA
Email: so.niknamian@gmail.com

Abstract

Reviewing and evaluating the effectiveness of advertising can be used as a criterion to convert weaknesses into strengths and budgeting. Also, Identification of the factors influencing the effectiveness of advertising can lead to more professional design and implementation. In the banking industry, with the increasing number of banks and increased right of customers to choose, the role of advertising has become increasingly important to succeed in this field of competition; so it should be said that advertising can make the bank close to its goals when it is sufficiently effective. The current research aimed to identify and rank the effective factors on the design of a native model for evaluation of the effectiveness of advertising of the Eghtesad Novin bank in Iran. The statistical population of this research is all managers and experts of the Eghtesad Novin bank. According to the research literature, the criteria and sub-criteria affecting the effectiveness of advertising have been extracted. By using the fuzzy Delphi technique and based on the experts' opinion, the final indices were extracted. Then, using the fuzzy AHP method, the rank of criteria has been addressed. The derived criteria are demographic characteristics, individual emotions, and attitudes of the customer, apparent features of the message, message content, market characteristics, cultural, economic features, design and nature of the service, advertising methods. The results of the ranking showed that the methods of advertisement, the design, and nature of the service, the apparent characteristics of the message, the emotions and attitudes of the customer, the demographic characteristics, the content of the message, economic, market characteristics, were placed in the first to ninth ranks, respectively.

Keywords: banking services, banking services marketing, advertising, advertising effectiveness,

Introduction:

Advertising is one of the promotional mix elements of marketing mixes, which play an important role in attracting customers. Advertising is a bridge between the organization and customers and can introduce the organization and its products to customers, inform consumers about the goods and services of the organization, convince them about the benefits of the product, and ensure them to have the best choice even after the purchase when the customer is uncertain (Buck, 2016: 172).

Given the increasing competition among the brands in the marketing sector, advertising is a necessary tool to introduce the characteristics of goods and services to customers. Leong et al. (1999) pointed out the fact that consumers overcome advertising information, and many of them do not remember the advertisements that they are exposed to, and it is necessary to consider the advertising and its attractive methods to attract customers' attention. The ability to attract customers' attention is one of the main indicators of advertising effectiveness that should be taken into consideration. The attention of customers is a key component of the overall effectiveness of advertising (Pileliene et al., 2017: 2). Any professional marketing without advertising will not be effective. Despite relatively high advertising costs, advertising is mentioned as an investment rather than the cost. Since many advertisements cause that the advertising companies do not benefit their advertising because of the wrong way of advertising, the competitive companies benefit from the advertising of that company (Hartnet, 2016: 23)

Advertising should be in line with the goals and plans, as well as the strategies of the organization. Therefore, we should see to what extent the advertising has met such goals. Allocating funds without evaluating the effectiveness of past advertising is the waste of resources of the organization. By evaluating the effectiveness of advertising, changes can be made to the form, the content of messages, type of media, budget, communication channels, and even the time and conditions for conducting advertising to create more useful and effective advertising than the past (Yazgirudo-Yasteh, 2015: 359). In the current competitive world, the importance of advertising is increasing. As a result, businesses have devoted a great deal of money to customers' attraction, but it is important to see whether the company advertising has been designed and implemented in such a way to meet the goals of the company. This issue has been often neglected in our country. (Kim, 2014: 2107).

The amazing effect of advertising on the lives of individuals requires that this scientific, social and communicative branch be carefully studied and advertising is made with a scientific, artistic, specialized and expert view. Professional advertisement can be a sharp sword for sales promotion, both inside and outside the country, and otherwise, it is a waste of time, money and energy. A professional advertising lever can, with the help of new technology, penetrate humans, and cause more desire for consumption (Arribeh, 2016: 463). Today, banks have turned their attention to the widespread use of mass media advertising, given the highly competitive structure governing on them. Also, given the relationship among a wide range of businesses of organizations and households and banks through various deposit accounts, loans, and other financial services, banks can benefit from advertising in a proper way to meet their goals, namely the introduction of their services, the communication with customers, their effectiveness on them in choosing a bank to do their banking business and ultimately attracting them. Unfortunately, in most banks, despite the massive advertising and the allocation of large amounts of energy and budget, the effectiveness of such advertisements is less evaluated. Since advertising is one-way, it is necessary to receive feedback from the customers (Rabiei, 2011: 19).

Regarding marketing, the effectiveness of advertising depends on the wide range of coverage of various consumers and the cost of advertising campaigns. Economists have defined the effectiveness of advertising as a change in the profit derived from promotional activities. Therefore, the effectiveness of bank advertising is the potential quality of information reported to consumers, and in promotional campaigns, it aims at creating a bank image and informing banking products and services, as well as obtaining a specific economic result before and after the transmission of the message (Denis et al., 2012: 85).

Eghtesad Novin Bank (EN bank), which is one of the oldest banks in Iran, due to the addition of new banks, should use new tools to protect and even increase its market share for the assessment of the effectiveness of its advertising. For this purpose and given that the bank uses various advertisements, it is necessary to design a model to evaluate the effectiveness of the bank advertising; therefore, the present study seeks to answer the following questions:

- What indicators are used to assess the bank native advertising model?

- How are the indicators for bank advertising evaluation weighed and ranked?

In this research, we have tried to identify the criteria using the Fuzzy Delphi process; then we will use the Fuzzy AHP method to rank the identified criteria. In the next section of the research, the literature has been studied. Then the research design has been expressed, and in the last section the conclusion has been presented.

1-Research literature:

1-1-Marketing of banking services

The growing trend in service activities has led to the emergence of new markets and large service companies. The complexity of service industries, such as information systems, management consultancy, and logistics, has increased, as service providers have entered the market with a variety of business backgrounds and capabilities to provide a broader range of services. Facing this complexity requires knowledge and skills to understand the services. The definition of the service includes all activities that have been done (for example, writing document specifications) (Selviaridis et al., 2013: 1398).

In the industrial literature, there are various ways to define the service and product requirements. Van Weele (2010) has distinguished technical and applied definitions from each other. The first requirement expresses the exact technical specifications of the goods with the required services and inputs while it focuses on performance demands and demand-oriented performance orientations. In this regard, Axelsson and Wynstra (2002) have proposed four methods for definition of services: input definitions are based on the sources and capabilities of the service provider while the process-oriented approach focuses on how services are presented regarding activities and processes. Output-oriented definitions focus on service capabilities and performance levels, as well as the interaction between the suppliers and recipients. Ultimately, they focus on stress-oriented results in the economic value of services (Selviaridis et al., 2013: 1399).

The financial sector forms an essential part of the service sector. In this regard, banks have used marketing to improve the strategic performance of banks. In particular, since the 1980s, this trend has been confirmed and criticized. Marketing researchers and experts are interested in researching the various aspects of marketing in banking. The quality of services is essential for customers who use banking services, and the provision of accurate and high-quality services leads to the survival of banks in the competitive market. Improving the quality of services will increase the access and response to the demands of the bank's customers, which will make the bank more distinct than other competitors. Therefore, measuring the quality of services in the banking industry is essential. Providing quality services is a hard and difficult task, in this regard, the features of services in general and banking services, in particular, should be explained. Banking services require a high degree of appropriate communication between service providers and customers. If services are not provided properly, they will be at high risk.

About the five features that distinguish them from material services (intangibility, non-allocation of materials for production, diversification, non-corruption, and lack of ownership), financial services have two other specific features, that have been firstly developed by McKechnie (1997). These characteristics include trusteeship and two-way information. These special features mean that the evaluation of services is more difficult than material products, and they not only create a challenge in the development of marketing activities, but also they are the source of the problem in providing an optimal level of quality services. Banking services require a high level of communication between service providers and customers, as well as a high degree of customer participation in the process of providing services. Angurr et al. (1999) emphasized that due to the high standardization and the development of technologies in the banking sector, human factor has been considered as the most important one in making distinctions among competitors. The expertise and motivation of the employees, as well as the level of customer involvement in the provision of services, significantly affect the quality of services. Due to the wide range of banking activities, attention should be paid to cultural and social changes that affect the attitudes of customers (Durdana et al., 2015: 100).

Marketing of banking services has three components as follows: if any of them were properly executed, banking services would change the country.

1- Market study: Identification of the banking services market system, including the status of current banking services and their weaknesses, strengths, expectations, demands, ideas and opinions of customers, the conditions of environmental instruments affecting the market system, in particular government policies and governing laws, and even global changes in the banking system and successful models in this system. Indeed, market study is the foundation of the evolution of the banking system, because market study (which is the result of constant communication with the market and conducting marketing studies and research in various forms) leads to comprehensive, up-to-date, relevant and necessary information for decision-makers and bank managers (Munsung Rheea & Satish Mehrab, 2006: 508).

2. Market creation: It involves the use of a variety of simple and complex marketing techniques and competitive strategies, as well as the growth, development and market creation in the banking system; it needs to understand the internal customers (employees) and external

customers, to communicate with them and meet their needs and demands. Market creation in the competitive environment requires appropriate strategies (Moch, 2013: 2908).

3. Market protection: One of the most important tasks of marketing in the banking system of the country is to maintain market share, a stable relationship with customers, and to convert ordinary customers to loyal and supportive ones. The more intense the competition, the more valuable the loyal customers will be, as marketing research in the world has shown that the cost of finding a new customer in firms is 3 to 33 times higher than the cost of maintaining current customers regarding the type of products and services. Therefore, market protection prevents the costly advertising, and attraction of new customers and provides a kind of confidence and serenity for market survival, especially in competitive conditions (Fredriksson & Moro, 2014: 69).

1-2-Effectiveness of bank advertising

One of the indicators of success in any organization is the extent to which the organization achieves its goals. Here, the success of any organization is determined by its own goals. This approach has been called effectiveness in the management literature. The effectiveness is the degree to which the organization achieves its goals. Goals must first be defined to measure effectiveness in any organization. Then, plans are set up to fulfill these goals, and thus the plans will be implemented. In the end, the results are compared with predetermined goals (Manjegani et al., 2015: 124).

About 100 years ago, William Heketh statement, "Half of the advertisements were wasted, we just do not know which half it is" attracted everyone to the accountability of advertising. What matters to advertisers is that how much the amount of money spent on advertising has returned? Has the cost spent on making and broadcasting the advertising been effective? Has the effect of advertising been strong or weak? Knowing the answer to these questions is important to the advertiser because today companies pay much money to make and broadcast advertising. According to Jack Trout, the money we use to advertise is very much. Every second of the World War II was $ 9,000, and Vietnam cost $ 22,000 per minute. But in the American Football Final, every minute of advertising cost $ 2 million, so the advertiser has the right to know about the effectiveness of his advertising (Aksoy, Atilla, 2005, Yeni Reklamcilik, Istanbul Bilgi Universitesi, 95). It is very difficult to measure the impact of long-term advertising on many occasions. Also, academically in the world of advertising, it is impossible to create a comprehensive theory that can be used to determine the effectiveness of advertisements on people at different levels under different circumstances. The effectiveness of media is its capacity to create desired results, or special conditions in a particular position and content, that is, attaining the desired communication goals (Wery, 2012)

Evaluating the effectiveness of advertising indicates how much advertising has succeeded in achieving the goals. The evaluation of the effectiveness of advertising acts as a feedback, informs us about the weaknesses and strengths of our advertising by which we find our shortcomings in the advertising. The marketer may choose an appropriate stage regarding product life or brand position, and use it to define his advertising goals and ultimately evaluate its effectiveness (Eckler & Bolis, 2011: 2).

The effectiveness of advertising is a function of its content (message), the execution (how it is transmitted), and repetition (the number of times each consumer sees the message) (Cutler, 1997, quoted by Moorthy & Hawkins, 2005). Some of the effectiveness criteria are the call and recognition of the brand name (Park, 2006), awareness, willingness and intention to purchase (Decrop, 2007), attitude towards the ad, attitude toward the brand name and intent to purchase (Muda et al., 2014)

The control of the communication element has a direct effect on customer perception of different dimensions of the brand name, such as service quality and loyalty. It means that the elements of communication, advertising and promotion must be carefully implemented and controlled (Suvittawat, 2015). Informative advertising is one of the factors that shape the behavior of the bank customers. Studies conducted both on attracting, retaining and loyalty of customers, show the need to examine the impact of advertising on attracting and loyalty of customers (Ioannou, Boukas and Skoufari, 2014: 25).

An ideal model of world-class advertising can: 1. transfer a consistent meaning to the consumer all over the world. 2. Transfer positive global capacity for the brand. That is, the results of advertising among different peoples should produce a positive effect and transfer high-quality signals (Strebinger et al., 2018: 410).

Advertisers want to prepare consumers to like goods or services, but sometimes it harms them through unwanted and intrusive advertising. It leads to the creation of a series of negative emotions in advertising. The purpose of ads is to create positive feelings in consumers about advertised goods and services (Bell & Buchner, 2018: 1).

On the other hand, research shows that commercial advertising can be positively evaluated and considered, and this positive evaluation leads to a positive evaluation of the brand with the product. However, advertising can have negative effects that lead to the formation of negative attitudes towards that product and service. Therefore, by increasing the volume of messages, it can have positive, negative or neutral effects. Therefore, marketers should evaluate advertising messages and their behavioral outcomes. With this regard, the effectiveness of advertising, based on a hierarchical approach, should be fully investigated (Tafani et al., 2018: 125).

There are various factors which influence effectiveness of advertisement like media, time, market life stage, advertisement life stage (new or old advertisement) and also advertisement creative signs (like content and administrative elements) (Talis et.al, 2005). Also using popular cartoon personages (Ranjbarian, 2009), size (Rousat, Venous & Ebrahimi, 2004, Baltas, 2003), color (Park, 2008, Moriarty, Mitchell, Wells, 2009, Cheng & Kao, 2011), complexity or simplicity of the advertisement (Morrison & Dainoff, 1972, Di Res, 2008), using irony factor (Mukherjee, Hober, 2004), using famous artists (Rousta, Venous, Ebrahimi , 2004) and consumer features (Matrin, Phaymi, Agee, 2002, Park, 2008) are among these variables. Advertisement repetition is also a factor which helps in reminding its content to the audiences (Rousta, Venous, Ebrahimi, 2004). One of considered factors in the debate of evaluating effectiveness of advertisement is the attitude towards advertisement. Attitude towards the advertisement includes public interest or lack of interest in an advertising stimulus in encountering a special topic (Moaen & Minor, 2009). Mental imagery (B one & Elen, 1992), content transparence and emotions called through the advertisement are among variables influencing the attitude towards the advertisement. Some of these factors may lead to strong impacts while some of them may create weak impacts or don't include any impact. Awareness of the impact of each of these factors makes a considerable help in making management decisions (Talis et.al, 2005).

Fotini et.al (2009) provided a model of effective advertisements for advertising websites of big companies. The results of the study showed that measurements, activities, priorities and intentions of people have a considerable impact on evaluation and qualitative factors of relation-oriented marketing have a great impact on advertisements effectiveness indicators. The extracted criteria include:

- Advertisement policies: product information, company information, product general display, consistency with traditional advertisement

- Medium features: message features, information structure, responding time, facilitating research, beauty, interactions, communications in real-time, amusement, information control

- Relation- oriented marketing policy: communications, feedback, supporting customers

Today, organizations for competition in the business world and selling their goods and services require purposeful and efficient marketing activities. One of topics which has a very important role in the process of marketing of an organization and assign a high volume of budget to it is advertisement. Advertisement causes some changes in the knowledge, attitude and behavior level of audiences, so attention to evaluating advertisement effectiveness is a necessity for the organizations that consider advertisement as requirement of continuous life, growth and development. By assessing advertisement effectiveness and investigating consistency with the organization objectives and its expected impacts, some changes may be created in its dimensions, content and methods so that a more effective and useful advertising is performed. Regarding the research literature and internal and external background which has been stated, here a general categorization of the model proposed criteria has been provided in the following table:

Table 1- summation and extracted variables

scholar	sub standard	criterion
) Dargi & Mohammadian (2010), Ashouri (2016), Mansouri Moayed (2005), Aghapour (2016), Mirzadh & Mehrmanesh (2016), Keramforn (2011), Mac Ki- Nesbiat et.al (2011), Rosemary et.al (2012), Wang (2015), Maryion et.al(2017)	awareness of sexuality differences	individual factors
Dargi & Mohammadian (2010), Ashouri (2016), Samsam Shariat Keramforn (2011)(2007), Mirzadeh & Mehrmanesh (2016), Mac Ki- Nesbiat et.al (2011), Rosemary et.al (2012), Wang (2015), Maryion et.al(2017)	attention to age requirements	
Dargi & Mohammadian (2010), Dehdashti Shahrokh (2009)	various consumers (permanent, ordinary and case consumers)	

Khashizadeh (2010), Danaei & Dasar (2016), Pillay et.al (2012), Marion et.al (2017)	advertisement richness	
Danaei & Dasar (2016), Dargi & Mohammadian (2010), Pillay et.al (2012), Olson et.al (2012)	advertisement accuracy	
Ashouri (2016), Rabiei (2011), Taeini Harari (2014)	positive mentality of the organization	
Dargi & Mohammadian (2010), Shahmohammadi (2011), Olson et.al (2012)	honesty feeling in advertisement	
Fakhari et.al (2012)	threatening feeling	
Ashourizadeh (2016), Hamidizadeh (2013)	attitude towards brand	
madhoushi (2013), Olson et.al (2012)	being convinced	
Ahmadi (2013), Mac Ki – Nesbiat et.al (2011), Olson et.al (2012)	convicting audiences	
Ashouri (2016), Hosivar (2014), Olson et.al (2012)	mental imagery	
Siadat & Sakhi Jourashk (2014)	positive feeling	
Navaei et.al (2015), Rozmari et.al (2012)	audience being active	
Mosavi Esmaeilzadeh (2015), Olson et.al (2012)	knowledge level	
Dargi & Mohammadian (2010), Ashouri (2016), Shafiei Nikabadi & Zarei (2017)	attitude towards advertisement	
Dargi & Mohammadian (2010), Hosseinzadeh et.al (2016), Sanei et.al (2016)	familiarity with methods and forms of advertisement	message
Ashouri (2016), Olson et.al (2012), Alan (2012), Patric Ro (2014)	attention to reduction of expectation time of the user for observing advertisement	
Danaei & Dasar (2016), Forentebury & Mac Gldric (2011)	beautiful and appropriate	

	combination of colors	
Aghapuor (2016), Sanei et.al (2016)	site and location of advertisement placement	
Hassanzadeh & Danaei (2016), Sanei et.al (2016)	access to sufficient and required information	
Dargi & Mohammadian (2010), Darzban (2016)	starting the advertisement with a challenge	
Navaei et.al (2015), Salimi Fard & Pourdehghan (2016), Bargar et.al (2016)	starting the advertisement with a statement and explanation	
Dargi & Mohammadian (2010), Mirzadeh & Mehrmanesh	using helping words and terms for decision making	
Dargi & Mohammadian (2010), Daneshvar et.al (2016), Alan (2012)	using stimuli with diversions more than usual and expected extent	
Sanei et.al (2016)	using images of women and children	
Sohrabi Ranani (2016), Fortenbury & Mac Goldric (2011)	creativity in the message	
Shafei Nikabadi & Zarei (2017), Denis et.al (2012)	the text of advertisement message being a predicate sentence	

Khashizadeh (2010), Dargi & Mohammadian (2010), Sohrabi Ranani (2016), Patric Ro (2014), Abdolsalam et.al (2016)	continuation and repetition of message	
Daneshfar & Nabizadeh (2015), Fortenbury & Mac Goldric (2011)	using trustable and famous people	
Salimifard & Pourdehghan (2016)	common dialects and terms	
Daneshfar & Nabizadeh(2015), Forentnburi & Mac Goldric (2011)	loveliness of the popular person for the observer	
Daneshfar & Nabizadeh (2015), Forentnburi & Mac Goldric)(2011	familiarity of famous person with the observer	
Mansouri Moayed (2005), Aghapour (2016)	emotional music	
Khonsazadeh (2010), Hosseinzadeh et.al (2016), Aghapour (2016)	using various methods	
Siadat & Sakhi Jourashk (2014), Sohrabi Ranani (2016)	high coverage and being universal	
Navaei et.al (2016), Mosavi Esmaeilzadeh (2015), Hosseinzadeh et.al (2016), Mirzadeh & Mehrmanesh (2016), Rosemary et.al (2012), Mikit et.al (2012), Spenkar (2013)	media advertisement	
Navaei et.al (2015), Mosavi Esmaeilzadeh (2015), Hosseinzadeh et.al (2016), Ashouri (2016), Hassanzadeh & Danaei (2016), Mikit et.al (2012)	multimedia advertisement	
Navaei et.al (2015), Mirzadeh & Mehrmanesh (2016)	radio advertisement	
Navaei et.al (2015), Mosavi Esmaeilzdeh (2015), Ashouri (2016), Hassanzadeh & Danaei (2016)	cell phone advertisement	
Navaei et.al (2015), Hosseinzadeh et.al (2016), Mirzadeh & Mehrmanesh (2016)	creating TV advertisement	
Ahmadi (2013), Salimi Fard & Pourdehghan (2016), Hoang et.al (2013)	attraction of rewards	
siadat & Sakhi Jourashk (2014), Koumra et.al (2016)	being benevolent	

Darzaian (2016), Zabkouich & Louksten (2011), Abdolsalam et.al (2016)	coordination in speaking and behavior	
Ahmadi (2013), Salimi Fard & Pourdehghan(2016), Honari et.al (2015)	reality of messages	
Madhoushi (2013), Salimi Fard & Pourdehghan (2016), Houang et.al (2013)	graphic-designing	
Shahmohammadi (2011), Madhoushi (2013), Salimi Fard & Pourdehghan(2016)	psychological – geographical division	
Mirzadeh & Mehrmanesh (2016), Zabkouich & Lokesten (2011), Menroz et.al (2012), Choubhai et.al (2013), Mokan & Oltz (2014), Ayono et.al (2014)	advertisement on web	
Dehdashti Shahrokh (2009), Shahmohammadi (2011), Mirzadeh & Mehrmanesh (2016), Menroz et.al (2012), Choubhai et.al (2013), Mokan & Oltz (2014), Ayono et.al (2014), Henri et.al (2015), Yazghiro- Basteh (2015), Lin (2016)	internet & cell phone advertisement	
Shahmohammadi (2011), Sanei et.al (2016), Menroz et.al (2012), Chobai et.al (2013), Ayono et.al (2014), Henri et.al (2015)	selecting advertising media	
Shahmohammadi (2011), Sanei et.al (2016), Howang et.al (2013), Henri et.al (2015), Yazghirdo-Yasteh (2015), Martina Santa et.al (2017)	artistic aspect of advertisement	
Kashizadeh (2010), Hoshivar(2014), Pelpin et.al (2017)	gender of speaker in the advertisement	
Hosseinzadeh et.al (2016), oribeh (2016), Madra et.al (2017)	text repetition	
Shahmohammadi (2011), Hassanzadeh & Danaei (2016), Abri & Joumar (2014), Ayouno et.al (2014), Oribeh (2016), Pachman (2016), Bak (2016), Madra et.al (2017)	advertisement	
Samsam Shariat (2007), Sohrabi Ranani (2016), Daneshvar et.al (2016), Bak (2016), Kiji-Kali (2017)	factors relating to usage	environment

Fakhari et.al (2012), Mirzadeh & Mehrmanesh (2016), Howang et.al (2013), Bak (2016), kiji-Kali (2017)	factors relating to facilitation	
Dargi & Mohammadian (2010), Hamidizadeh (2013), Darzban (2016), Belman et.al (2012)	awareness of different status of the environment status	
Dargi & Mohammadian (2010), Mirzadeh & Mehrmanesh (2016)	awareness of access location of objective group to the product	
Dargi & Mohammadian (2010), Salimifard & Pourdehghan (2016)	awareness of norms	
Dargi & Mohammadian (2010), Hamidizadeh (2013), Darzban (2016), Belman et.al (2012), Koumra et.al (2016)	awareness of religions and religious values	
Dargi & Mohammadian (2010), Hassanzadeh & Danaei (2016)	awareness of method of evaluating services in relation to other services	
Hassazadeh & Danaei (2016), Radmiker (2015)	attention to security	
Samsam Shariat (2007), Mirzadeh & Mehrmanesh (2016), Belman et.al (2012)	economic beliefs	
Salimifard & Pourdehghan (2016), Pilay et.al (2012), Howang et.al (2013), Komra et.al (2016)	ethical beliefs	
Salimifard & Pourdehghan (2016)	Islamic teaching	
Samsam Shariat (2007), Daneshvar et.al (2016), Radmiker (2015)	strategic look to advertisement plan	
Dargi & Mohamadian (2010), Ansari & Rasei (2016), Abri & Joumar (2014)	market share	
Dargi & Mohamadian (2010), Darzain (2016)	obviating primary needs of audiences	

Salimi Fard & Pourdehghan (2016), Kim et.al (2012), Howang et.al (2013)	reference groups	
Sohrabi Ranin (2016), Fontini et.al (2009), Belman et.al (2012)	awareness of services	factors relating to services
Salimi Fard & Pourdehghan (2016), Abri & Joumar (2014), Radmiker (2015)	stating potential benefits of purchase ad services consumption	
Dargi & Mohamadian (2010), Hosseinzadeh et.al (2016), Fontini et.al (2009)	rate of services complexity	
Samsam Shariat (2007), Darzaian (2016), Fontini et.al (2009), Radmiker (2015)	display of expected experience	
Khashizadeh (2010), Darzain (2016)	rate of familiarity of consumer with services	
Ansari & Rasaei (2016), Belman et.al (2012)	market researches	
Hosseinzadeh et.al (2016), Mirzadeh & Mehrmanesh (2016)	record of exploiting services	
Dargi & Mohammadian (2010), Ansari & Rasaei (2016), Lousi (2012), Park (2015), Marion et.al (2017)	being unique	
Dehdashti Shahrokh (2009), Daneshvar et.al (2016), Radmiker (2015), Jankoski et.al (2016)	price, benefits	
Daneshvar et.al (2016), Fontini et.al (2009), Jankoski et.al (2016)	relation with customer	
Samsam Shariat (2007), Daneshvar et.al (201), Fontini et.al (2009), Jankoski (2016)	supporting customer	
Dehdashti Shahrokh (2009), Ansari & Rasaei (2016), Fontini et.al (2009), Radmiker (2015), Jankoski (2016)	feedback	

2- Research plan

The present study is of descriptive applied type which has been performed with survey method. The statistical population of this study includes all managers and experts of Eghtesad

Novin Bank. The statistical population of the present study contains all people who are exposed to advertisements of Eghtesad Novin Bank across the country. For selecting samples, Cochran formula is used for unlimited societies that based on this formula, the samples number were determined 400 persons for each province. The sample people were selected based on the following sampling method. At first, using stratified sampling method , the samples were randomly selected from five geographical regions of the country (north, south, east, west and center) of nine provinces (Mazandaran, Khorasan Razavi, Zanjan, Chahar Mahal Bakhtiari , Kerman, East Azarbaijan , Kermanshah, Ilam and Tehran).

3- Variables and research model

In this study, fuzzy Delphi technique has been used for identifying main indicators. For ranking the final criteria and sub standards, multi-criteria decision making models and AHP model have been used with fuzzy approach. For analysis of obtained data, Excel software and coding in the environment of this software have been used. The results obtained from data analysis have been provided in the present chapter.

3-1 Identifying criteria and sub standards of factors effective on advertisement effectiveness

At the first step, the primary indicators were identified and selected. In the first stage, by reviewing the literature and performed studies, a large number of indicators were extracted from the literature. For screening the indicators and identifying the final indicators, fuzzy Delphi approach has been used. The expert views have been collected in the field of rate of significance of main indicators. Fuzzy average and defuzzification output of rates relating to indicators are presented in table 2. The rate of defuzzification greater than 7 is acceptable and every indicator which has the score higher than 7 is confirmed (Wu& Fang, 2011).

Table 2- fuzzy average and fuzzy screening of indicators of factors effective on advertisement effectiveness

Result	Crisp	mean	U	M	L	
acceptance	7.15	(2, 7.97, 9)	9	7.97	2	Q01
acceptance	7.84	(6, 8.01, 9)	9	8.01	6	Q02

Result	Crisp	mean	U	M	L	
acceptance	7.92	(6, 8.13, 9)	9	8.13	6	Q03
acceptance	7.39	(4, 7.83, 9)	9	7.83	4	Q04
acceptance	7.70	(6, 7.8, 9)	9	7.80	6	Q05
rejection	6.68	(2, 7.28, 9)	9	7.28	2	Q06
acceptance	7.71	(6, 7.82, 9)	9	7.82	6	Q07
acceptance	7.66	(6, 7.75, 9)	9	7.75	6	Q08
acceptance	7.02	(2, 7.78, 9)	9	7.78	2	Q09
acceptance	7.27	(2, 8.15, 9)	9	8.15	2	Q10
acceptance	7.09	(2, 7.88, 9)	9	7.88	2	Q11
rejection	6.85	(2, 7.52, 9)	9	7.52	2	Q12
acceptance	7.43	(4, 7.9, 9)	9	7.90	4	Q13
rejection	6.71	(3, 7.07, 9)	9	7.07	3	Q14
acceptance	7.17	(2, 8.01, 9)	9	8.01	2	Q15
acceptance	7.13	(2, 7.95, 9)	9	7.95	2	Q16
rejection	6.85	(4, 7.02, 9)	9	7.02	4	Q17
acceptance	7.43	(4, 7.89, 9)	9	7.89	4	Q18
acceptance	7.70	(6, 7.8, 9)	9	7.80	6	Q19
rejection	6.80	(3, 7.21, 9)	9	7.21	3	Q20
acceptance	7.75	(6, 7.88, 9)	9	7.88	6	Q21
rejection	6.62	(3, 6.93, 9)	9	6.93	3	Q22
acceptance	7.61	(6, 7.66, 9)	9	7.66	6	Q23
acceptance	7.22	(2, 8.08, 9)	9	8.08	2	Q24
acceptance	7.04	(2, 7.81, 9)	9	7.81	2	Q25
acceptance	7.71	(6, 7.82, 9)	9	7.82	6	Q26
rejection	6.65	(3, 6.97, 9)	9	6.97	3	Q27
acceptance	7.98	(6, 8.22, 9)	9	8.22	6	Q28
acceptance	7.70	(6, 7.79, 9)	9	7.79	6	Q29
rejection	6.54	(3, 6.81, 9)	9	6.81	3	Q30
acceptance	7.19	(4, 7.53, 9)	9	7.53	4	Q31
acceptance	7.79	(5, 8.18, 9)	9	8.18	5	Q32

Result	Crisp	mean	U	M	L	
acceptance	7.53	(4, 8.05, 9)	9	8.05	4	Q33
acceptance	7.05	(2, 7.82, 9)	9	7.82	2	Q34
acceptance	7.43	(4, 7.89, 9)	9	7.89	4	Q35
rejection	6.92	(3, 7.38, 9)	9	7.38	3	Q36
acceptance	7.02	(2, 7.78, 9)	9	7.78	2	Q37
rejection	6.91	(3, 7.37, 9)	9	7.37	3	Q38
rejection	6.96	(2, 7.68, 9)	9	7.68	2	Q39
rejection	6.99	(2, 7.74, 9)	9	7.74	2	Q40
acceptance	7.42	(4, 7.88, 9)	9	7.88	4	Q41
acceptance	7.10	(2, 7.9, 9)	9	7.90	2	Q42
rejection	6.59	(2, 7.13, 9)	9	7.13	2	Q43
rejection	6.55	(2, 7.08, 9)	9	7.08	2	Q44
acceptance	7.77	(6, 7.9, 9)	9	7.90	6	Q45
rejection	6.91	(2, 7.62, 9)	9	7.62	2	Q46
acceptance	7.09	(2, 7.88, 9)	9	7.88	2	Q47
acceptance	7.68	(6, 7.78, 9)	9	7.78	6	Q48
acceptance	8.19	(6, 8.53, 9)	9	8.53	6	Q49
rejection	6.88	(2, 7.57, 9)	9	7.57	2	Q50
acceptance	7.75	(5, 8.12, 9)	9	8.12	5	Q51
acceptance	7.03	(2, 7.8, 9)	9	7.80	2	Q52
rejection	6.96	(4, 7.19, 9)	9	7.19	4	Q53
acceptance	7.04	(2, 7.8, 9)	9	7.80	2	Q54
rejection	6.96	(3, 7.45, 9)	9	7.45	3	Q55
acceptance	7.02	(2, 7.78, 9)	9	7.78	2	Q56
acceptance	7.92	(5, 8.38, 9)	9	8.38	5	Q57
acceptance	7.00	(2, 7.76, 9)	9	7.76	2	Q58
rejection	6.90	(1, 7.86, 9)	9	7.86	1	Q59
acceptance	7.10	(4, 7.4, 9)	9	7.40	4	Q60
acceptance	7.86	(6, 8.04, 9)	9	8.04	6	Q61
rejection	6.97	(2, 7.71, 9)	9	7.71	2	Q62

Result	Crisp	mean	U	M	L	
acceptance	7.02	(2, 7.79, 9)	9	7.79	2	Q63
acceptance	7.73	(6, 7.84, 9)	9	7.84	6	Q64
acceptance	7.01	(2, 7.76, 9)	9	7.76	2	Q65
rejection	6.65	(3, 6.98, 9)	9	6.98	3	Q66
acceptance	7.01	(2, 7.77, 9)	9	7.77	2	Q67
acceptance	7.90	(6, 8.11, 9)	9	8.11	6	Q68
acceptance	7.55	(5, 7.83, 9)	9	7.83	5	Q69
acceptance	8.02	(6, 8.28, 9)	9	8.28	6	Q70
acceptance	7.95	(6, 8.17, 9)	9	8.17	6	Q71
acceptance	7.03	(2, 7.79, 9)	9	7.79	2	Q72
acceptance	7.65	(5, 7.98, 9)	9	7.98	5	Q73
acceptance	7.00	(1, 8, 9)	9	8.00	1	Q74
acceptance	7.04	(2, 7.8, 9)	9	7.80	2	Q75
acceptance	7.03	(2, 7.8, 9)	9	7.80	2	Q76
acceptance	7.30	(4, 7.7, 9)	9	7.70	4	Q77
rejection	6.88	(2, 7.57, 9)	9	7.57	2	Q78
acceptance	7.63	(4, 8.19, 9)	9	8.19	4	Q79
acceptance	7.65	(5, 7.98, 9)	9	7.98	5	Q80
rejection	6.59	(3, 6.88, 9)	9	6.88	3	Q81
acceptance	7.09	(2, 7.88, 9)	9	7.88	2	Q82
rejection	6.83	(2, 7.5, 9)	9	7.50	2	Q83
acceptance	7.12	(2, 7.93, 9)	9	7.93	2	Q84

From the 84 indicators which were assessed, 24 indicators which have acquired the average score less than 7 are omitted. So, 60 indicators are finally confirmed (Adelazaret.al, 1995).

The remained indicators will be finally examined in the format of a 60-question questionnaire with analytical hierarchy process.

3-2 determining the priority of factors effective on effectiveness of advertisement using fuzzy AHP technique

For determining the priority of major criteria, fuzzy analytical hierarchy (FAHP) has been used. The analysis process is as follows.

1- Dual comparison of major criteria based on objective and determining the weight of major criteria

2- Dual comparison of sub standards of each criterion and determining sub standards of each cluster

3- Multiplication of the weight of sub standards in the weight of respective criterion and determining the final weight of sub standards

For dual comparison of elements, 9-degree Saaty scale has been used. 9-degree Saaty scale has been provided by Thomas Saaty , codifier of analytical hierarchy theory. Also, in this study fuzzy approach has been used for making the values quantitative. So, Saaty fuzzy range has been used.

Table 3- fuzzy range equal to 9-degree Saaty scale in AHP technique. Habibi et.al , 2014, 77

reversed fuzzy equivalent	fuzzy equivalent	the word term of comparison status of i to j
(1,1,1)	(1, 1, 1)	Preferred Equally
$\left(\frac{1}{3},\frac{1}{2},1\right)$	(1, 2, 3)	intermediate
$\left(\frac{1}{4},\frac{1}{3},\frac{1}{2}\right)$	(2, 3, 4)	Preferred moderately
$\left(\frac{1}{5},\frac{1}{4},\frac{1}{3}\right)$	(3, 4, 5)	intermediate
$\left(\frac{1}{6},\frac{1}{5},\frac{1}{4}\right)$	(4, 5, 6)	Preferred Strongly
$\left(\frac{1}{7},\frac{1}{6},\frac{1}{5}\right)$	(5, 6, 7)	intermediate
$\left(\frac{1}{8},\frac{1}{7},\frac{1}{6}\right)$	(6, 7, 8)	very strongly Preferred
$\left(\frac{1}{9},\frac{1}{8},\frac{1}{7}\right)$	(7, 8, 9)	intermediate
$\left(\frac{1}{9},\frac{1}{9},\frac{1}{9}\right)$	(9, 9, 9)	Extremely Preferred

(source in papersL Khorshid & Zabihi, Lee my H.I; Wen-Chin Chen, Ching-Jan, Chang, 2008, 101)

The pattern of criteria and sub standards has been drawn in the figure using AHP technique. Also, the criteria and sub standards of the study have been named with numeric index so that the research flow is easily detectable.

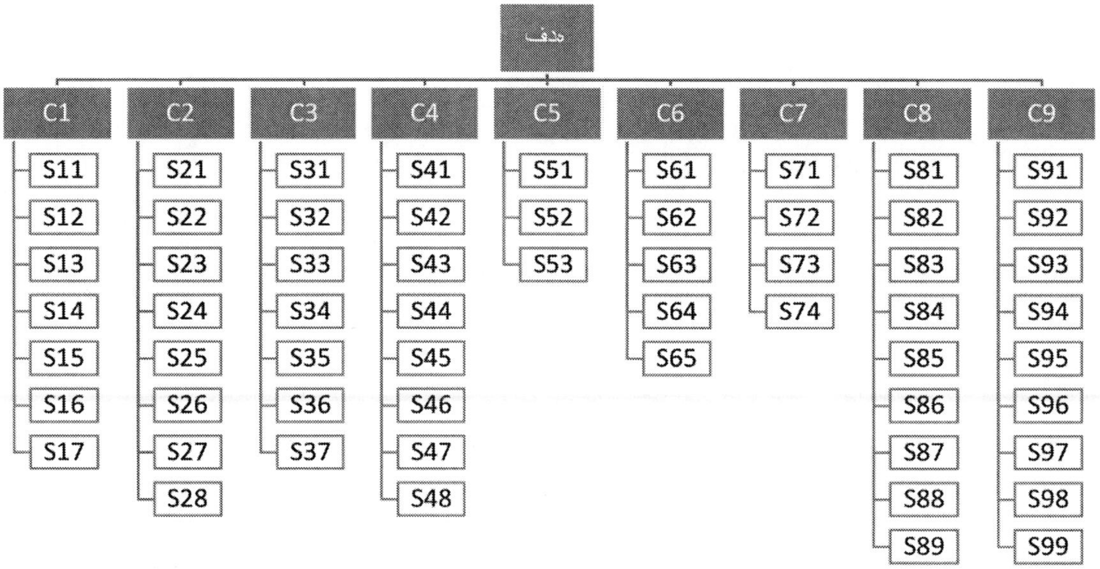

Figure 1- hierarchical model of criteria and sub standards of factor effective on advertisement effectiveness

1- Determining the priority of major criteria based on objective

In the first step, the major criteria based on the objective have been compared dually. Dual comparison is very simple and all elements of each cluster should be compared two by two. Therefore, if there is n element in a cluster, $\frac{n(n-1)}{2}$ comparisons will be performed. Since there are four criteria, so the number of performed comparisons is equal to:

$$\frac{n(n-1)}{2} = \frac{9(9-1)}{2} = 36$$

So, 36 dual comparisons have been performed from group perspective of experts. The experts view has been made quantitative using fuzzy scale. Collecting experts view has been collected with 9-degree Saaty range. So, 6 dual scales of four major four criteria have been performed based on the experts view. Using geometric average the experts views have been aggregated. Fuzzy aggregation will be calculated as follows:

Equation 1

$$F_{AGR} = (\prod(l), \prod(m), \prod(u))$$

After matrix formation of the obtained dual comparisons, the special vector has been calculated. At first, the fuzzy extension of each row is calculated. Every entry of dual comparison matrix $\tilde{X}$

Is displayed as $\tilde{x}_{ij}$.

The fuzzy extension of every row is also displayed with symbol $\tilde{S}_i$. Therefore, the fuzzy extension of each row is calculated as follows:

Equation 2

$$\tilde{S}_i = \sum_{j=1}^{n} x_{ij}$$

Then, the fuzzy extension of every row elements will be as follows:

Fuzzy extension of row 1 (1,1,1) ⊕ (1,27,1,47,1,75) ⊕ (0.47, 0.55, 0.66) ⊕ (1.52, 1.81, 2.14) ⊕ (0.93, 1.14, 1.39) ⊕ (1.26, 1.61, 2.04) ⊕(1.25, 1.51, 1.84) ⊕(0.78, 0.97, 1.24) ⊕ (0.59, 0.75, 0.96) ⊕ (9.08, 10.8, 13.02)

Fuzzy extension of row 2 (0.57, 0.68, 0.79)⊕(1, 1, 1)⊕(0.58, 0.7, 0.89)⊕(1.84, 2.28, 2.75)⊕(0.68, 0.82, 1.02)⊕(1.42, 1.81, 2.34)⊕(1.64, 2.04, 2.53)⊕(0.71, 0.88, 1.12)⊕(0.73,
fuzzy extension of row 3 (1.52, 1.82, 2.11)⊕(1.13,)0.9, 1.1)⊕ = (9.17, 11.11, 13.55
1.42, 1.72)⊕(1, 1, 1)⊕(1.89, 2.34, 2.88)⊕(0.7, 0.86, 1.09)⊕(1.45, 1.78, 2.22)⊕(1.55, 1.88, 2.29)⊕(0.57, 0.71, 0.93)⊕(0.67, 0.8, 0.96)⊕ = (10.47, 12.61, 15.2)

Fuzzy extension of row 4 (0.47, 0.55, 0.66)⊕(0.36, 0.44, 0.54)⊕(0.35, 0.43, 0.53)⊕(1, 1, 1)⊕(0.97, 1.21, 1.51)⊕(0.96, 1.33, 1.8)⊕(2.14, 2.7, 3.37)⊕(0.94, 1.21, 1.56)⊕(1.3, 1.69, 2.16)⊕ = (8.49, 10.57, 13.14)

Fuzzy extension of row 5 (0.72, 0.88, 1.07) ⊕ (0.98, 1.22, 1.48)⊕(0.92, 1.17, 1.43)⊕(0.66, 0.82, 1.03)⊕(1, 1, 1)⊕(0.67, 0.83, 1.04)⊕(0.23, 0.29, 0.36)⊕(0.28, 0.32, 0.39)⊕(0.3, 0.38, 0.48)⊕ = (5.76, 6.91, 8.29)

Fuzzy extension of row 6 (0.49, 0.62, 0.79)⊕(0.43, 0.55, 0.7)⊕(0.45, 0.56, 0.69)⊕(0.55, 0.75, 1.04)⊕(0.96, 1.2, 1.48)⊕(1, 1, 1)⊕(0.34, 0.41, 0.51)⊕(0.27, 0.34, 0.44)⊕(0.24, 0.31, 0.41)⊕ = (4.74, 5.75, 7.08)

Fuzzy extension of row 7 (0.54, 0.66, 0.8)⊕(0.4, 0.49, 0.61)⊕(0.44, 0.53, 0.64)⊕(0.3, 0.37, 0.47)⊕(2.81, 3.5, 4.3)⊕(1.95, 2.43, 2.96)⊕(1, 1, 1)⊕(0.34, 0.42, 0.54)⊕(0.29, 0.36, 0.47)⊕ = (8.06, 9.76, 11.78)

Fuzzy extension of row 8 (0.8, 1.03, 1.28)⊕(0.89, 1.14, 1.41)⊕(1.08, 1.41, 1.76)⊕(0.64, 0.82, 1.06)⊕(2.55, 3.1, 3.63)⊕(2.25, 2.97, 3.72)⊕(1.86, 2.39, 2.94)⊕(1, 1, 1)⊕(0.58, 0.69, 0.83)⊕ = (11.66, 14.56, 17.64)

Fuzzy extension of row 9 (1.04, 1.33, 1.69)⊕(0.91, 1.11, 1.37)⊕(1.04, 1.25, 1.5)⊕(0.46, 0.59, 0.77)⊕(2.08, 2.63, 3.3)⊕(2.42, 3.23, 4.1)⊕(2.15, 2.74, 3.43)⊕(1.2, 1.46, 1.72)⊕(1, 1,)1)⊕ = (12.29, 15.35, 18.88

Then, the fuzzy sum of total priorities column elements is calculated

Equation 3

$$\sum \tilde{S}_i = \sum_{i=1}^{n} \sum_{j=1}^{n} x_{ij}$$

The sum of elements of the priorities column of major criteria will be as follows:

$$\sum_{i=1}^{4} \sum_{j=1}^{4} M_g^j = (79.72, 97.41, 118.59)$$

For normalization of each criterion priorities, that criterion total rates sum should be divided into sum of all priorities (column elements), since the rates are fuzzy, so fuzzy sum of every row is multiplied in summation reverse. The summation reverse should be calculated.

Equation 4

$$if\ \tilde{F} = (l, m, u)\ then\ \tilde{F}^{-1} = \left(\frac{1}{u}, \frac{1}{m}, \frac{1}{l}\right)$$

Then, based on the equation 4, we will have

$$\left(\sum_{i=1}^{n} \sum_{j=1}^{n} M_g^j\right)^{-1} = (0.01, 0.01, 0.01)$$

$$P_k = \sum_{i=1}^{n} M * \left(\sum_{i=1}^{n} \sum_{j=1}^{n} M_g^j\right)^{-1}$$

So, the results obtained from normalization of obtained rates will be as follows:

$\widetilde{W}_{C1} =$ (0.08, 0.11, 0.16)

$\widetilde{W}_{C2} =$ (0.08, 0.11, 0.17)

$\widetilde{W}_{C3} =$ (0.09, 0.13, 0.19)

$\widetilde{W}_{C4} =$ (0.07, 0.11, 0.16)

$\widetilde{W}_{C5} =$ (0.05, 0.07, 0.1)

$\widetilde{W}_{C6} =$ (0.04, 0.06, 0.09)

$\widetilde{W}_{C7} =$ (0.07, 0.1, 0.15)

$\widetilde{W}_{C8} =$ (0.1, 0.15, 0.22)

$\widetilde{W}_{C9} =$ (0.1, 0.16, 0.24)

Each of the obtained rates is fuzzy and normalized weight relating to major criteria. There are various methods for defuzzifcation of the obtained rates. In this study, Bojadziev proposed method has been used for defuzzification.

Table 4- defuzzification and computation of final and normal weight of criteria of factors effective on advertisement effectiveness

Normal	Deffuzy	X3max	X2max	X1max	Crisp
0/111	0/117	0/113	0/115	0/117	C1
0/114	0/120	0/115	0/118	0/120	C2
0/103	0/137	0/133	0/135	0/137	C3
0/108	0/113	0/112	0/113	0/113	C4
0/070	0/073	0/072	0/073	0/073	C5
0/060	0/063	0/062	0/063	0/063	C6
0.101	0/107	0/103	0/105	0/107	C7
0.149	0/157	0/153	0/155	0/157	C8
0.158	0/167	0/163	0/165	0/167	C9

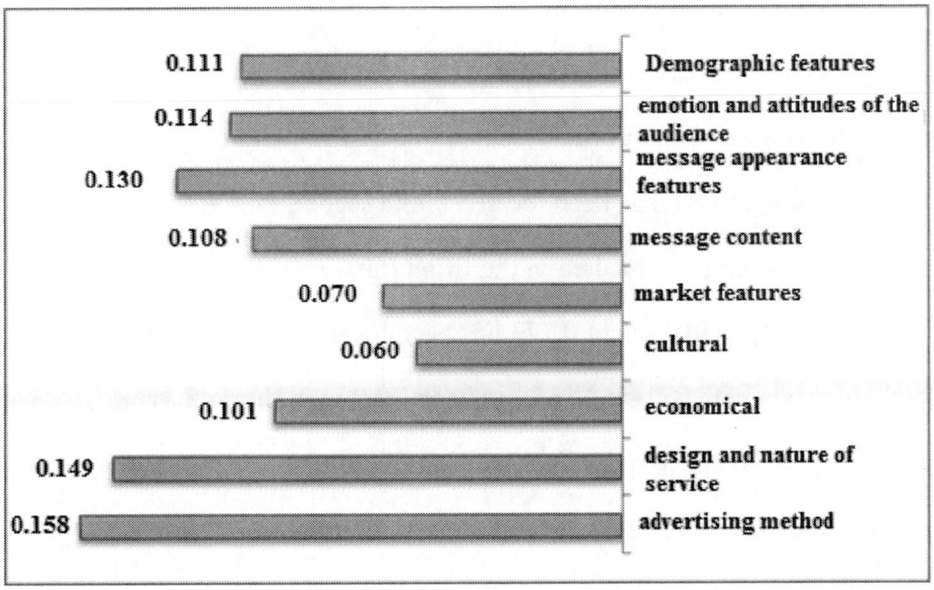

Figure 2- graphical display of significance of criteria of factors effective on advertisement effectiveness

Based on the obtained special vector:

- Demographic features with weight 0.111 have been located in rank five

- Individual feelings and attitude of the addressee with weight 0.114 has been located in rank four.

- The message surface features with weight 0.13 has been located in rank three.

- The message content with weight 0.108 has been located in rank six

- The market feature with weight 0.07 has been located in rank eight.

- Cultural with weight 0.06 has been located in rank nine.

- Economic with weight 0.10 has been located in rank seven.

- Plan and nature of services with weight 0.149 has been located in rank second.

- Advertisement methods with weight 0.158 has been located in rank first.

The inconsistency rate of performed comparisons has been obtained 0.030 which is smaller than 0.1 and so the performed comparisons is sufficed.

2- Comparison and determining priority of sub standards

At first, the experts' views have been collected in the dual comparison of sub standards. Then, the performed calculations for fuzzification of the experts views average have been provided for determining the priority of sub standards.

Priority of sub standards of demographic features

The audience job with weight 0.151 has been located in the third rank.

The audience income with weight 0.093 has been located in the seventh rank.

The audience religion with weight 0.16 has been located in the second rank.

Social class with weight 0.141 has been located in the fifth rank.

Awareness of gender differences with weight 0.173 has been located in the first rank.

The individual knowledge level with weight 0.135 has been located in the sixth rank.

Priority of sub standards of the audience individual feeling and attitude

Positive mentality with weight 0.124 has been located in the fifth rank.

Feeling of honesty in advertisements with weight 0.089 has been located in the eighth rank.

Threatening feeling with weight 0.14 has been located in the third rank.

Mental image with weight 0.102 has been located in the seventh rank.

Attitude towards brand with weight 0.121 has been located in the sixth rank.

Convincing audiences with weight 0.149 has been located in the first rank.

Attitude towards advertisement with weight 0.143 has been located in the second rank.

The individual perception of accuracy of advertisement with weight 0.133 has been located in the fourth rank.

Priority of sub standards of the message surface features

Then, based on the results of performed dual comparisons:

Beautiful and appropriate combination of colors with weight 0.143 has been located in the fourth rank.

Attention to reduction of the user waiting time with weight 0.159 has been located in the second rank.

Graphics- design with weight 0.131 has been located in the sixth rank.

Creativity in the message with weight 0.137 has been located in the fifth rank.

Emotional music with weight 0.099 has been located in the seventh rank.

Text repetition with weight 0.153 has been located in the third rank.

Access to required and sufficient information with weight 0.178 has been located in the first rank.

Determining the priority of sub standards of the message content

Benevolent with weight 0.093 has been located in the eighth rank.

Coordination in speech and behavior with weight 0.125 has been located in the fourth rank.

Reality of the messages with weight 0.119 has been located in the sixth rank.

Using women and children images with weight 0.131 has been located in the third rank.

Gender of speaker in the advertisement with weight 0.106 has been located in the seventh rank.

Using trustful and famous people with weight 0.16 has been located in the first rank.

Loveliness of famous people for the observer with weight 0.122 has been located in the fifth rank.

Familiarity of the famous person with the observer with weight 0.144 has been located in the second rank.

Priority of sub standards of the market features

Then, based on the results of the performed dual comparisons:

Factors relating to its application in the market with weight 0.326 have been located in the second rank.

Awareness of different aspects of status and environment with weight 0.283 has been located in the third rank.

Awareness of evaluation method of service receiver with weight 0.391 has been located in the first rank.

Priority of cultural sub standards

Then, based on the results of performed dual comparisons:

Awareness of norms with weight 0.199 has been located in the third rank.

Awareness of religion and religious values with weight 0.238 has been located in the first rank.

Ethical beliefs with weight 0.18 have been located in the fourth rank.

Obviating primary needs of the audiences with weight 0.164 has been located in the fifth rank.

Reference groups with weight 0.219 have been located in the second rank.

Determining the priority of economic sub standards

Then, based on the results of the performed dual comparison:

Economic beliefs with weight 0.261 have been located in the second rank.

Strategic look at an advertising program with weight 0.215 has been located in the fourth rank.

Market share with weight 0.274 has been located in the first rank.

Attention to economic security with weight 0.251 has been located in the third rank.

Priority of the plan and services nature sub standard

Awareness of services with weight 0.104 has been located in seventh rank.

Expressing potential benefits of purchase and services consumption with weight 0.117 has been located in the third rank.

Services complexity rate with weight 0.089 has been located in the ninth rank.

Uniqueness with weight 0.108 has been located in the sixth rank.

Rate of familiarity of the services consumer with weight 0.12 has been located in the second rank.

Price and allowances with weight 0.114 has been located in the fourth rank.

Relation with the customer with weight 0.111 has been located in the fifth rank.

Displaying the expected experience with weight 0.098 has been located in the eighth rank.

Supporting customers with weight 0.139 has been located in the first rank.

Priority of sub standards of the advertising methods

Beginning the advertisement with a challenge with weight 0.079 has been located in the ninth rank.

Beginning an advertisement with a statement and explanation with weight 0.107 has been located in the fifth rank.

Using stimuli with diversions more than usual and expected with weight 0.097 has been located in the seventh rank.

The advertisement message text being predicating with weight 0.126 has been located in the third rank.

Using diverse methods with weight 0.088 has been located in the eighth rank.

Media advertisement with weight 0.113 has been located in the fourth rank.

Dimensions of TV advertisement with weight 0.151 have been located in the first rank.

Psychological, geographical division with weight 0.107 has been located in the fifth rank.

Advertising media selection with weight 0.132 has been located in the second rank.

Determining the final priority of major criteria with FAHP technique

For determining final priority of major criteria using FAHP technique, the weights relating to major criteria (W1) and weight of indicators based on each criteria (W2) are needed. The results of comparison of the research sub criteria relating to them forms matrix W2. For determining the final priority of indicators with AHP technique, it is sufficient that the weight of indicators based on every criterion (W2) is multiplied in the major criteria (W1). Any of these matrices has been calculated in the previous steps. Based on the obtained output, it is obvious that the market share item, one of economic factors with weight 0. 277 is the most important indicator in the existing indicators. Item of awareness of evaluation method of service receiver in relation to other services with weight of 0.0273 has the second priority. Item of economic believes, one of economic factors, has the third priority.

So, the research model is as follows:

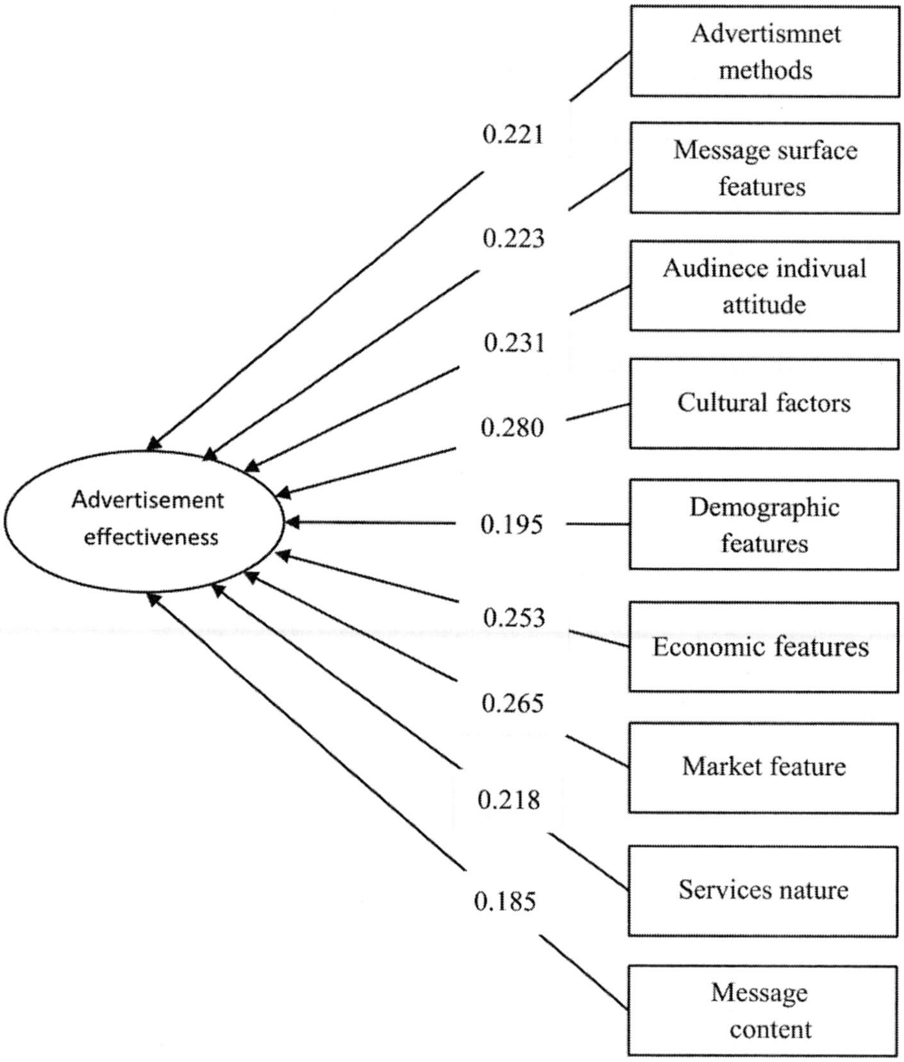

4- Conclusion:

In this study, identification and ranking of factors effective on designing the native model of advertisement in Eghtesad Novin bank has been addressed. In this first step, the primary indicators were identified and selected. In the first stage, by reviewing the literature and performed studies, a great number of indicators were extracted from the literature. For screening the indicators and identifying final indicators, fuzzy Delphi approach has been used. The experts' views have been collected in the field of significant rate of major indicators. From 84 indicators which were assessed, 24 indicators which have acquired score less than 7 are omitted. So, 60 indicators were finally confirmed. Then, regarding the performed calculations, the final weight of any of the model indicators have been calculated with fuzzy AHP. Based on the obtained output, it is obvious that the market share of economic factors with weight of 0.027 is the most important indicator in the existing indicators. The item of awareness of evaluating services in relation to other services is one of market features factors with weight 0.0273 which has second priority. The item of economic beliefs with weight 0.0264 is in the third priority.

Today, considering extension of competition in the international level and formation of world competition, all countries of the world may be successful in this arena to coordinate them with this trend and create more value for their customers by services and products they provide. In the today modern world, almost all people are exposed to advertisements and other communicational activities. Effectiveness of advertisement is one of important issues in the advertisement domain. In this regard, Mansouri Moyed (2005) showed that attention attraction, commercial advertisements of Maskan Bank are effective but in respect of other variables, that is, creating interest, stimulation of tendency and measurement to purchase, commercial advertisements of Maskan Bank are not effective. Samsam Shariat (2007) showed that TV channels and cartoon animations , attractive package, colors relating to the product and innovation, noticing safety feeling and associated symbols, informing of quality, background and price, dialects and common phrases, diversity in the message content, in TV, broadcasting from 18-23 and before beginning programs (in media) have been effective factors in effectiveness of commercial effectiveness from people viewpoint. Mosavi Esmaeilzadeh (2015) showed that oral advertisements (in 4 dimensions) and after that TV advertisement (just in 2 dimensions) are effective. Also, radio, internet and billboard advertisement are not effective. And also Friedman test was used for ranking advertisement tools. Salimi Fard & Pourdehghan (2016) showed that among factors effective on attitude to cell phone advertisement, components of general attitude towards advertisement, knowledge level and value have respectively the greatest significance and components of time of presentation and reference groups with the least significance in effecting the individuals attitude towards cell phone advertisement. Ansari & Rasaei (2016) showed that the creativity of advertisement message , market research, competition, market share, uniqueness and relation with customer are effective on the effectiveness of brand advertisement. Louis (2012) showed that advertising in the urban public spaces will have a great impact in the consumers' mentality. Since some advertisements are transitory and are often displayed in a short period of time and just in several spaces, but advertisement in public spaces exist everywhere and is similar to advertising through print and media, but since they have lower cost for production and distribution, have lower cost in this respect. Zabckowich & Loukston (2011) showed that in necktie advertisements use more of sensory requests and information signs. Findings showed that in western and European countries, people establish more contact due to globalization increase. They declared that content of printed advertisements, includes visual and text advertisements which are effective in establishing relation with many audiences. Also, ethnic identity as an important cultural variable increases attention to consumer purchase behavior. Besides that, they found out that the image of people, models, ceremonies and managers constitute an important part of images and background image plays an important role in effectiveness of visual factors in advertisement. Mac Ki- Nesbiat et.al (2011) showed that younger adults remind emotional messages specially negative ones better than others and older adults more prefer positive and logical messages to negative and emotional ones, but attitude towards advertisement among younger adults is not different. Also, they found that older adults have the most positive attitude towards advertisements.

Perzouti (2015) showed that advertising models are different in various ages and each age is consistent with a special model and using various and consistent models with people conditions may make the advertisement more effective. Abri & Joumar (2014) showed that advertising messages are effective on the consumer behavior and the advertiser may use various media for communicating consumer and media schedule is effective on the advertisement success. Attention to environmental factors where the customer lives are important factors in the advertisement success. Madra et.al (2017) showed that the advertisement text, text repetition impact the advertisement effectiveness. Maryion et.al (2017) showed that simultaneous use of several media like TV and internet may increase effectiveness of the advertisement and considering two variables of gender and multimedia create a better individual attitude towards the message effectiveness.

In ranking the identified factors, advertisement methods have assigned the first rank among factors. It is suggested that in designing advertisement and selecting advertising instruments, the dimensions of TV advertising, selecting median and predicative message text should be noticed. Using advertising instrument like internet advertisement, adverting on public vehicles like buses and metro due to great number of people who use them can increase stability of the advertisements in the customers mind.

In codifying and designing bank advertisements, the design and services nature should be noticed. It is proposed that supporting customer, familiarity of consumer with services are considered and during the advertisement process, it is required that the potential benefits of purchase and services consumption are expressed.

In designing the advertisement and content of bank advertising message, it is required to use famous and trustful people for introducing the services. Also, children as future customers and audiences should be noticed since they are future potentials. Total advertisements provided by the bank through various media should move in a unit line so that the audience can establish a logical relation between the observed advertisements in various media that is total advertising efforts lead to synergy of advertisement effectiveness. This issue won't be acquired unless through a media planning and efficient advertisement.

In advertisements, the message surface features should be noticed. It is required that in advertisement, access to required and enough information for the customers, attention to reduction of expectation time of user for observing the advertisement and suitable repetition of the text are considered. Providing large information in an advertisement which has a short time and providing information that require special mathematical analysis should be avoided.

References:

Hosseinabadi, Maryam, Mahmoud Nematian, Mohammad Besharati, investigating the relation between mixed elements of marketing and customers satisfaction with Tehran

Sarmayeh bank branches, management and entrepreneurship studies, fall 2016, 2nd period, No. 3 (page 162-169)

Dargi Parviz, 2008, subjects and topics of marketing management with Iran market attitude, Tehran, Rasa publication

Sarmad, Zohreh, Bazargan Abbas, Hejazi, Elaheh, 2014, research methods in behavioral sciences, Tehran, Agah. 27-13

Salimi Fard, Khodakaram, Pourdehghan, Adel, 2016, identifying and ranking factors effective on attitude towards cell phone advertisement using AHP, Brand management, period 3, no. 6, summer 2016, page 133-160

Seyed Akbar Manjgani, Vahideh Tabasi Lotfabadi, 2015, factors effective on effectiveness of advertisement on financial stoop of legal clients of Mashhad City with AHP approach, finance journal, fall 2015, No. 75 (page 121-146)

Sanayei, Ali, Majid Mohammad Shafiei, Maedeh Amini Valashani, 2016, effectiveness of bank advertisement in customer attraction, commercial management searches, spring and summer 2016, No. 15 (page 185-210)

Mahmoudian, Mahmoud, Rabiei, Ali, Jamili, Bita. 2011, evaluation of effectiveness of Parsian bank advertisement and identifying the most important factors in increasing its effectiveness in Tehran city, modern marketing studies, summer 2011, No. 2 (page 17-40).

Manjagani, Seyed Akbar, Tabasi Lotfabadi, Vahideh, 2015, factors effective on advertisement effectiveness on financial stoop of legal clinets of Mashhad city with AHP approach, finance journal, fall 2015, No. 75, 1-121

Venous, Davar, Safaeian, Mitra, (2005), marketing bank services in Iran and bank services features, Tehran, Negah Danesh, 4th edition

Angulo, N. (2016). 10 digital marketing trends to watch in 2016 and beyond. Retrieved from: http://www.marketingdive.com/news/10-digital-marketing-trends-to-watch-in-2016-./and-beyond/411457

Bell Raoul, Buchner, Axel.(2018). Positive Effects of Disruptive Advertising on Consumer Preferences, Journal of Interactive Marketing 41 (2018) 1–13

Corvi. Elisabetta.(2010).The effectiveness of advertising: a literature review, 10th Global Conference on Business & Economics, ISBN : 978-0-9830452-1-2

Cramphorn Michael F.,(2011), Forum Gender effects in advertising, International Journal of .Market Research Society Vol. 53 Issue 2

Cronin, J. J. & Fox , G. L. (2010). "The implications of third party customer complaining for .advertising efforts". Journal of advertising. 39(2): 21-23

Decrop, A. (2007). The influence of message format on the effectiveness of print advertisements for tourism destinations. International Journal of Advertising, 26(4), 505–.525

Denis .A. Shakhov and ,Anna A. Panasenko.(2012).Evaluating Effectiveness of Bank Advertising in the Internet: Theory and Practice, World Applied Sciences Journal 18 (Special Issue of Economics): 83-90, 2012 ISSN 1818-4952

Denis 1,2 A. Shakhov and 3Anna A. Panasenko.(2012).Evaluating Effectiveness of Bank Advertising in the Internet: Theory and Practice, World Applied Sciences Journal 18 (Special Issue of Economics): 83-90, 2012

Durdana Ozretic-Dosen Ines Zizak , (2015),"Measuring the quality of banking services targeting student population", EuroMed Journal of Business, Vol. 10 Iss 1 pp. 98 - 117

Eckler, P., and Bolls, P. (2011). Spreading the Virus: Emotional Tone of Viral Advertising and Its Effect on Forwarding Intention and Attitudes. Journal of Interactive Advertising, .11(2): 1-11

Eric W.T. Ngai Ka-leung Karen Moon S.S. Lam Eric S. K. Chin Spencer S.C. Tao , (2015),"Social media models, technologies, and applications", Industrial Management & Data Systems, Vol. 115 Iss 5 pp. 769 – 802

Fotini Patsioura Maro Vlachopoulou Vicky Manthou, (2009),"A new advertising effectiveness model for corporate advertising web sites", Benchmarking: An International Journal, Vol. 16 Iss 3 pp. 372 - 386

Frederic Crane, , G. (2010). «Marketing FO Entrepreneurs: ConceptandApplication for new .Venture. New York: SAGE Pablication, Inc

Frolova. Svetlana.(2014).THE ROLE OF ADVERTISING IN PROMOTING A PRODUCT, Thesis CENTRIA UNIVERSITY OF APPLIED SCIENCES ,Degree Programme in Industrial Management May 2014, p"1-61

Henry Boateng Abednego Feehi Okoe , (2015),"Consumers' attitude towards social media advertising and their behavioural response", Journal of Research in Interactive Marketing, Vol. 9 Iss 4 pp. 299 -312

Henry Boateng Abednego Feehi Okoe , (2015),"Consumers' attitude towards social media advertising and their behavioural response", Journal of Research in Interactive Marketing, Vol. 9 Iss 4 pp. 299 -312

Huang. Jinsong, Song Sub, , Liuning Zhouc, , , Xi Liu .(2013).Attitude Toward the Viral Ad: Expanding Traditional Advertising Models to Interactive Advertising, Journal of Interactive Marketing, Volume 27, Issue 1, February 2013, Pages 36–46

Idris, I., M.S.A. Yajid and A. Khatibi, 2009. Personal and Psychological Factors-Does it .Impact the Choice of Advertising Medium? Journal of Social Sciences, 5(2): 104-111

Ioannou Myria, Nikolaos Boukas, Evi Skoufari .(2014).Examining the role of advertising on the behaviour of co-operative bank consumers, Journal of Co-operative Organization and Management, Volume 2, Issue 1, June 2014, Pages 24-33

Jankowski. Jarosław, Przemysław Kazienko, Jarosław Wątróbski, Anna Lewandowska, Magdalena Zioło.(2016).Fuzzy multi-objective modeling of effectiveness and user experience in online advertising, Expert Systems with Applications, Volume 65, 15 December 2016, Pages 315-331

Kim, C., Park, S., Kwon, K., & Chang, W. (2012). An empirical test to measure the effectiveness of online advertising in online marketplaces using a hierarchical Bayes model. .Expert Systems with Applications (39), 117–128

Kim, ji-Hern., Hyun, yong j., (2010). " A model to investigate the influence of marketing – mix efforts and corporate image on brand equity in the IT software sector " . industrial . Marketing Management. No 40,2011, pp 424-438, Emerald Group Publishing Limited

Kim, S., & Jun, J. (2016). The impact of event advertising on attitudes and visit intentions. .Journal of Hospitality and Tourism Management, 29, 1-8

Kirby, J. and Marsden .(2006). Connected marketing : the viral, buzz and word of mouth .revolution, Oxford : Butterworth-Heinemann, 2006

*Madera. Quetzali, Oscar Castillo *, Mario García-Valdez, Alejandra Mancilla..(2017). A method based on Interactive Evolutionary Computation and fuzzy logic for increasing the effectiveness of advertising campaigns, Information Sciences 414 (2017) 175–186*

Marion Garaus and Udo Wagner , Anna-Maria B¨ack.(2017). The Effect of Media Multitasking on Advertising Message Effectiveness, Psychology & Marketing, Vol. 34(2):)138–156 (February 2017

Martín-Santana J.D., E. Reinares-Lara , P. Reinares-Lara .(2017).Influence of radio spokesperson gender and vocal pitch on advertising effectiveness: The role of listener gender, Spanish Journal of Marketing – ESIC, Volume 21, Issue 1, February 2017, Pages 63-71

McKay-Nesbitt, J., V. Manchanda, R., C. Smith, M., & A. Huhmann, B. (2011). Effects of age, need for cognition, and affective intensity on advertising effectiveness. Journal of Business Research (64), 12–17

Moch, Nils (2013). Competition in fragmented markets: New evidence from the German banking industry in the light of the subprime crisis. Journal of Banking &; Finance, Volume 37, Issue 8, Pp 2908–2919

Moorthy, S. and Hawkins, S. (2005). Advertising repetition and quality perception. Journal of .Business Research, 58, 354– 360

Mucan. Burcu, Özeltürkay. Eda Yaşa.(2014).Social Media Creates Competitive Advantages: How Turkish Banks Use This Power? A Content Analysis of Turkish Banks through Their Webpages, Procedia - Social and Behavioral Sciences, Volume 148, 25 August 2014, Pages 137-145

Muda, M.; Musa, R.; Niana-Mohamed, R. and Borhan, H. (2014). Celebrity Entrepreneur .Endorsement and Advertising Effectiveness. Social and Behavioral Sciences, 130, 11-20

Oladele. Kehinde Joseph, Ogunnaike, Akinbode Mosunmola, Aka Deborah .(2016). Effective Advertising: Tool For Achieving Client-Customer Relationships, Researchjournali's Journal of Media Studies Vol. 2 | No. 1 February | 2016

Olsen, G., Pracejus, J. W., & O'Guinn, T. (2012). Print advertising: White space. Journal of .Business Research (65), 855–860

Park, M. (2006). The compensatory effects of pictorial and verbal information for haptic information on consumer responses in non-store shopping environments. Unpublished Ph.D, .Ohio, Ohio state University

Park, T. (2008). Effective design of advertisement in handheld devices (Doctoral dissertation .)PURDUE UNIVERSITY

Park, T. (2008). Effectiveness Design of Advertisement in Handheld Devices. Dissertation for .doctoral degree, Purdue University, West Lafyette India

Pergelova, A., D. Prior and J. Rialp, 2010 Assessing Advertising Efficiency Does the Internet Play a Role? Journal of Advertising, 39(3): 39-54

Pilelienė Lina Viktorija Grigaliūnaitė , (2017)," The effect of female celebrity spokesperson in FMCG advertising: neuromarketing approach ", Journal of Consumer Marketing, Vol. 34 - .Iss 3 pp

Rosemary J. Averya, Matthew D. Eisenbergb, Kosali I. Simon .(2012).The impact of direct-to-consumer television and magazine advertising on antidepressant use, Journal of Health Economics, Volume 31, Issue 5, September 2012, Pages 705–718

Selviaridis. Kostas, Martin Spring b, Luis Araujo b .(2013). Provider involvement in business service definition: A typology, Industrial Marketing Management 42 (2013) 1398–1410

Strebinger .Andreas, Xiaoling Guo, Ferdinand Klauser, Peter Grant-Hayd .(2018).Is Multi-Ethnic Advertising a globally viable strategy for a Western luxury car brand? A mixed-method cross-cultural study, Journal of Business Research 82 (2018) 409–416

Suvittawat, A., (2015). "Commercial Bank Service Quality: Chonburi Province, Thailand". .International Journal of Management Sciences and Business Research. 4(2): 2226-8235

Tafani. Eric, Elyette Rouxa, Rainer Greifeneder .(2018).In the mood for action: When negative program-induced mood improves the behavioral effectiveness of TV commercials, Journal of Business Research 84 (2018) 125–140

Te'eni-Harari, T. (2014). "Clarifying the relationship between involvement variables and .advertising effectiveness among young people". Journal of Consum Policy. 37: 183–203

Thomas, S. and Kohli, S.C. (2011). Can Brand Image Move Upwards after Sideways? A .Strategic Approach to Brand Placements. Business Horizons, 54 (1): 41-49

Yahyaoui . Hamdi, Own .Hala S..(2018). Unsupervised clustering of service performance behaviors, Information Sciences 422 (2018) 558–571

Yuksel, A., Yuksel, F. & Bilim, Y. (2010) "Destination Attachment: Effects on Customer Satisfaction and Cognitive, Affective and Cognitive Loyalty", Journal Of Tourism Management, Vol.31, pp: 274-284.

STUDY OF NATURAL ANTIOXIDANT INHIBITORS FOR CORROSION PROTECTION OF MILD STEEL IN ACIDIC MEDIUM

Ms. Ibtihal Hassan AL-Bahrani, National University of Science and Technology, Sultanate of Oman.

ABSTRACT

Corrosion is a natural phenomena and it is a very big loss for structure as well as business; so corrosion minimization is one of the biggest areas of research. In this project; corrosion inhibition efficiency of berries juice extract was tested in mild steel sample against the sulfuric acid medium which was prepared from four types of berries; blackberries, blueberries, cranberries, and raspberries.

The experiment was designed by using Design Expert V 11 where immersed time, acid and berries juice extract concentrations were set as factors and a limited range was selected. The corrosion rate was set as a response. 20 runs were under taken and the experiments were done in the lab with maximum care.

The corrosion rate of 1020.374 mpy minimized to 597.659 mpy in presence of 5ml of extract and the efficiency of berries juice extract reached to 41.427%. The RSM gave 232.022 mpy corrosion rate as a minimum value and the optimum equation given by RSM was used for Genetic Algorithm analysis which gave 188.866 mpy corrosion rate at the optimizing conditions.

The surface and chemical characterization of the mild steel sample in presence and absence of berries juice extract were analyzed along with the relations between temperature, immersed time, acid and berries juice extract concentrations with response. The mixing tank was designed, calculations and drawing were presented.

KEYWORDS

Corrosion, Mild steel, Berries, Inhibitor, Weight loss, Sulfuric acid, RSM, SEM.

INTRODUCTION

Mild steel is used in most of our daily life applications as well as in industrial applications. In other hands, mild steel has low corrosion resistance which corrodes easily once it reacts with the surrounding environment; corrosion is a phenomenon where the mild steel reacts with the deteriorates and surrounding environment. Corrosion of mild steel is a continuous and constant problem where it severe financial and life loses as it leads to loss in the product, efficiency of equipment and failures in some cases (Roberge,2012).

The corrosion phenomenon controlled by adding a small amount of component to the corrosive environment which is a corrosion inhibitor. There are organic and inorganic inhibitors. The inorganic inhibitors are used in limit amount as they are toxic such as nitrites, molybate, nitrate, silicate, chromate, and phosphate (Cafferty,2010). Now a day, the corrosion engineers looking for cheap, renewable, friendly, readily available and ecologically acceptable inhibitors which are organic inhibitors which comes from the plants such as antioxidant components consisting of tannins, saponins, flavonoids, alkaloids and amino acids. Organic inhibitors consist of some heteroatoms such as oxygen, nitrogen, multi-bound, and sulphur atoms with the conjugated system which allows the absorption in the metal surface. Those atoms attract the absorption of some components on the metal surface and that done by the interaction of metals and by the electron of the conjugated system or by forming a bond with the free electron of heteroatom which forms a layer in the metal surface, as a result, it decreases the corrosion rate (Roberge,2012).

Berries are small pulpy fruits which have up to 9.2 mmol of antioxidant per 100g of berries. Berries can be used as a corrosion inhibitor as they are rich in antioxidant. The studies show that the berries juice extract of blueberries, cranberries, raspberries, and blackberries rich with antioxidant where the antioxidant capacity of them reach to 10 mg/ 100 g of blueberries (Michalska&Lysiak,2015), 58 mg/100g of cranberries (Blumberg,J.,et al,2013), 26.2mg/ 100 g of raspberries (Bobinaite,R.,Viskelis&Venskutonis,2015), and 21 mg/100 g of blackberries (Kaume,L.,et al,2012).

OBJECTIVES

To prepare anticorrosive extract (berries), to study corrosion in sample in presences and absences of the inhibitors, to determine the optimum conditions to inhibit corrosion and to design mixing tank.

METHADOLOGY, DESIGN AND EXPERIMENTATION

PREPARATION OF MILD STEEL SAMPLES

The mild steel sheets were cut into 2.5*3 cm^2 by using steel cutter and cleaned by surface cleaning machine to remove any corroded products in the metal surface then 4mm hole done by using drilling machine. Bunches for all metal samples (1-20) were made to differentiate between them. The drilled samples were washed by distilled water then rinsed with benzene and acetone respectively and dried with a hot air dryer. They were weighted using sensitive balance and kept in the desiccator until the time of usage.

PREPARATION OF CORROSIVE MEDIUM

The corrosive medium of different concentrations of H_2SO_4 prepared by using 36.8 N of the sulfuric acid.

PREPARATION OF BERRIES JUICE EXTRACT

137g, 209g, 282g, and 200g of raspberries, cranberries, blueberries, and blackberries respectively have been used to produce 309.5 ml of berries juice extract. The different types of berries washed by cold water, crashed in the electrical crasher and compressed mechanically by hand then filtered by cloth filter. The filtrate filtered again by using filter paper. The berries juice extract kept in the freezer at 0°C till the time of usage.

The efficiency of berries juice extract has been tested by running three experiments. The cleaned and weighted mild steel samples immersed in 0ml(blank), 2ml and 5ml of berries juice extract respectively with 50 ml of 1M H_2SO_4. After 2 hours; the samples have been taken out and cleaned by distilled water, benzene and acetone then dried by air dryer and weighted again. The corrosion rate and inhibitor concentration have been calculated by using equations (1) and (2) respectively.

$$Cr(mpy) = 22300 \times \frac{\Delta W}{\rho \times A \times t} \quad \textbf{equation(1)}$$

$$\mathrm{IE}(\%) = \frac{\text{initalcorrosionrate-finalcorrosionrate}}{\text{initalcorrosionrate}} \times 100 \quad \textbf{equation(2)}$$

DESIGN THE EXPERIMENT CONDITIONS

The experiments conditions have been designed by using Design Expert software-version 11 and following Response Surface Module(RSM); the range of the parameters have been given as inputs to the software which are inhibitor concentration (5-15 ml), immersed time (180-240 min) and acid concentration (1-2 M) and the response set as corrosion rate in mpy. Table(1) shows the experiment conditions given by the software.

Table(1): Designed runs given by RSM.

Run.No	Factor(A): inhibitor, ml	Factor(B): Time, min	Factor(C): Acid concentration, M
1	10	180	1
2	15	240	2
3	5	120	1
4	5	240	1
5	10	180	1.5
6	15	180	2
7	10	120	1.5
8	5	120	1.5
9	10	180	2
10	10	180	1.5
11	10	180	1.5
12	15	240	1
13	15	120	1
14	5	240	2
15	10	180	1.5
16	15	180	1.5
17	5	120	2
18	10	180	1.5
19	10	180	1.5
20	10	240	1.5

APPLYING THE DESIGNED RUNS IN THE LABORATORY

The weighted mild steel sample has been hanged by using a plastic thread in a 50 ml of sulfuric acid in presence of berries juice extract of known concentration and contact time as per the software; the batch has been kept in the shaker. After the contact time the mild steel sample cleaned by using 120 grade emery paper to remove any corroded products, washed by distilled water, rinsed by benzene and acetone respectively and dried by hot air dryer, weighted by using sensitive balance. The corrosion rate in mpy was calculated by using equation(1) where the weight losses results have been used.

STUDYING THE IMPACT OF TEMPERATURE IN CORROSION RATE

To study the impact of temperature in the corrosion rate; mild steel samples weighted and hanged for 2 hours in 50 ml of 1M sulfuric acid with 10ml of berries extract at different temperatures(30, 40, 50 and 60 °C) as shown in Table(2); each batch has been kept in the water bath shaker with heater to provide perfect contact between the solution and mild steel sample and to maintain the required temperature. Corrosion rate has been calculated by equation(1).

Table(2): Experiment conditions of temperature into corrosion rate.

SI.NO	Temperature (°C)	Inhibitor concentration (ml)	Acid concentration (M)	immersed time, (hours)
1	30	10	1	2
2	40	10	1	2
3	50	10	1	2
4	60	10	1	2

SURFACE AND CHEMICAL CHARACTERIZATION ANALYSIS

The mild steel surface of blank, absence and presence of berries juice extract samples have been studied by using Scanning Electron Microscopy (SEM) and Energy Dispersive Spectroscopy (EDS).

OPTIMIZATION STUDY BY GENETIC ALGORITHM

Matlab has been used for optimization where the genetic algorithm method has been followed. The actual equation form Design expert(RSM) has been used where "Optimtool" followed. The same limit of the conditions from RSM has been selected. The same previous procedure has been followed in the lab with maximum care by using those optimum conditions which given by the software, 1.001M of H_2SO_4 with 15 ml of berries juice extract immersed for 237.208 minutes.

DESIGN OF MIXING TANK

Design of mixing tank is a part of the project where the industrial scale mixing tank has been designed, the tank will be formed from high resistance material such as S.S to be able to resist the corrosive medium of sulfuric acid which will be mixed with berries juice extract. It is a container which used to mix various material (heterogeneous materials) together to form homogenous materials even the different viscosity materials can be mixed easily which provides continuous process, good control, easy cleaning, and simplest construction. It can be made from S.S material which provides a smooth surface and it can be easily cleaned (Petropedia,2018).

Parameters to be considered to design mixing tank are length, torque, weight, stiffness and hydraulic loading.(Dynamixianc,2015). The important factors in design mixing tank are tank type, tank volume, specific gravity, process and environment.(Ibryamoval,Y.et al,2009)

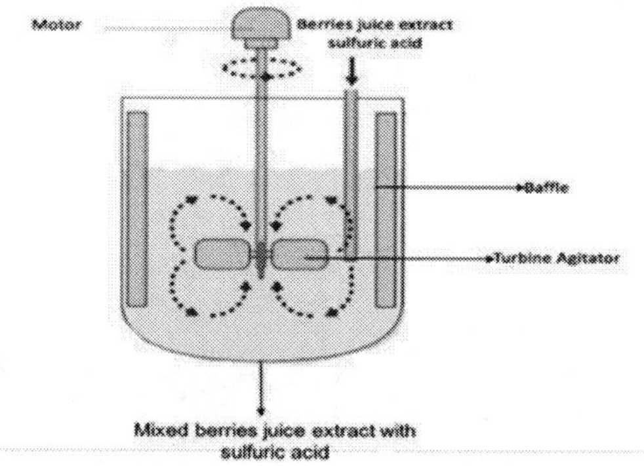

Figure(1): Mixing Tank(*Mixing Tank,*2018)

From the graph in Figure(2) Turbine agitator has been selected as it is suitable with the viscosity of the fluid and the tank volume. (Sinnot&Towler,2009)

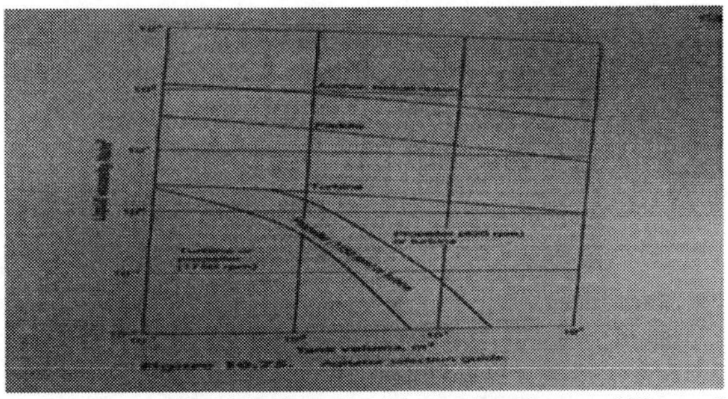

Figure(2): Graph of liquid viscosity Vs tank volume. (Sinnot&Towler,2009)

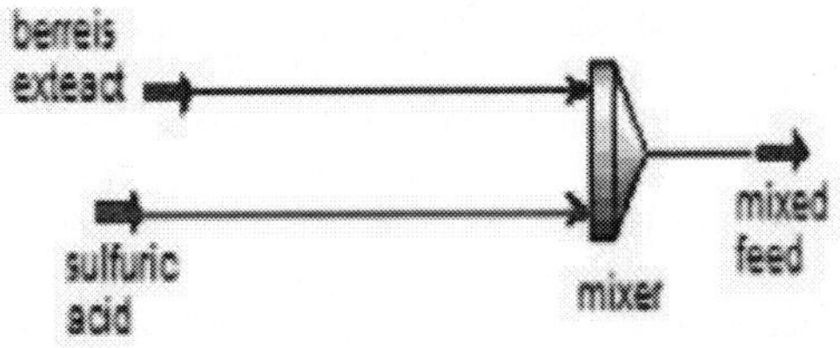

Figure(3): Result of Mixing Tank by Aspen Hysys.

Table(3): Calculated parameters for designing mixing tank

Tank diameter	m	3.5
Tank Height	m	1.81
Across sectional area	m^2	9.62
Volume of the tank	m^3	17.5
Impeller diameter	m	1.16
Length of paddle	m	0.29
Width of paddle	m	0.23
power	watt	204085
Reynolds Number	-	136486.6391

RESULTS AND ANALYSIS

EFFECT OF BERRIES JUICE EXTRACT CONCENTRATION INTO CORROSION RATE AND INHIBITOR EFFICIENCY

As shown in Table(4) the corrosion rate in the absence of berries juice extract inhibitor reach to the maximum (1020.374mpy); the corrosion rate has been minimized to 898.117mpy and 597.659mpy in presence of 2 and 5 ml of berries juice extract inhibitor respectively which means that berries juice extract inhibits the corrosion as the concentration of the berries juice extract increases. As well as the inhibitor efficiency reaches to 41.427% which obtained at 5ml of berries juice extract inhibitor in 1M of sulfuric acid immersed for 2 hours. The adsorption of the organic compounds such as ascorbic acid that present in the berries juice extract is blocking the active sites on the mild steel samples as it is forming a protective film on the metal surface as what kumar.A,2012 mention in his studies so the corrosion rate keeps decreasing as the amount of the berries juice extract is increasing. (Kumar.A,2012) Those compounds have heteroatoms such as oxygen which is a center of adsorption. In addition; It forms a coordinate bond as it has the ability to transfer the electrons to the vacant d-orbital of mild steel from the heteroatoms. (Ostovari, L. et al,2009)

Table(4): Experimental results of testing the efficiency of berries juice extract.

SI.NO	Inhibitor concentration (ml)	Corrosion rate (mpy)	Inhibitor efficiency (%)
1	0	1020.374	-
2	2	898.117	11.976
3	5	597.659	41.427

EFFECT OF IMMERSED TIME INTO CORROSION RATE AND INHIBITOR EFFICIENCY

The maximum corrosion rate reaches to 765.005mpy which obtained at 5ml of berries juice extract in the presence of 50 ml of 2 M of H_2SO_4 at 2 hours. Since the minimum corrosion rate of 232.022mpy obtained in presence of 5ml of berries juice extract with 50 ml of 1 M of H_2SO_4 immersed for 4 hours. The effect of immersed time into corrosion rate can be observed at low acid concentration and in the presence of berries juice extract, the corrosion rate is minimized at high immersed time. The decrease in the corrosion rate with the increase in immersed time is related to the adsorption of the corrosion product ($FeSO_4$) as shown in equation(3); According to Xianghong,L. et al ; the corrosion product of $FeSO_4$ becomes thicker with the immersed time, so it form a protective film (barrier) in the surface of mild steel (Xianghong, L. et al,2009).

$$Fe_{(s)}+H_2SO_{4(aq)} \rightarrow FeSO_{4(aq)} + H_{2(g)}$$, **equation(3)**

EFFECT OF ACID CONCENTRATION INTO CORROSION RATE

The effect of acid concentration into corrosion rate has been studied by using 1, 1.5 and 2 M of H_2SO_4 at different inhibitor concentrations and immersed time where the corrosion rate increase with the acid concentration at low inhibitor concentration and immersed time which reach to 765.005mpy at 2M H_2SO_4, 5ml of berries juice extract which immersed for 120 minutes as a maximum corrosion rate as shown in Table (3). In presence of the corrosive medium of H_2SO_4 with less amount of berries juice extract; the mild steel surface was exposed to H_2SO_4 medium where the corrosion reaction started as shown in equation(3) and $FeSO_4$ was formed with H_2 bubbles.

EFFECT OF TEMPERATURE INTO CORROSION RATE

From Table(5) can be noted that the rate of corrosion increases with temperature where it reaches to maximum 2407.994mpy at maximum temperature of 60°C. According to Ambrish, S. et al, at high temperature the berries juice extract molecules were partially absorbed in the mild steel surface so it was exposed to the corrosive medium of H_2SO_4, as a result, increasing the corrosion rate (Ambrish,S. et al,2012).

Table(5): Results of studying the effect of temperature into corrosion rate

SI.NO	Temperature,° C	Corrosion rate, mpy
1	30	701.507
2	40	1666.404
3	50	2008.861
4	60	2407.994

ANALYSIS BY RSM

Once the experimental results and corrosion rate calculated the RSM analysis has been done where the corrosion rate has been given to the software as a response. The response must be modified where the results fit with linear. The fit summary shows that the model F-value and P-value of 12.66 and 0.0001 respectively imply that the model is significant where A, B, C, and B^2 are the significant terms in this model only. 1.48 is a lack of fit P-value which indicates that the fit is not significant as there are pure errors.

0.5562 of the predicted R^2 is a reasonable agreement with 0.7105 of the adjusted R^2. The signal to noise ratio of 13.7252 of Adeq precision that the ratio grater than 4 is desirable. (Mycourses,2016)

The process variables are studied using ANOVA. The estimate coefficients in term of coded factors are A-Inhibitor, B-time, C-Acid and B^2. It is creating an accompanying condition for the sample with 1.0000 correlation coefficient. The coded and actual equations given by software are shown in equation(4) and (5) respectively.

- Corrosion rate(mpy)= 370.52-40.52(A)-126.97(B)+ 80.85(C) +66.58$(B)^2$ **equation(4)**
- Corrosion rate(mpy)= 1189.15487-8.10479(Inhibitor) -8.77421(Time) +161.69302 (Acid Conc)+0.018495 $(\text{Time})^2$ **equation(5)**

The various relations can be studied from one factor graph; it can be noted that the corrosion rate increases with acid concentration and decreases with the increase in the amount of berries juice extract and immersed time as can be seen from the plots in Figure(3):Plot pf corrsoion rate Vs inhibitor concntration, Figure(4): plot of corrosion rate Vs acid concentration and Figure(5):Plot of corrosion rate Vs immersed time respectively.

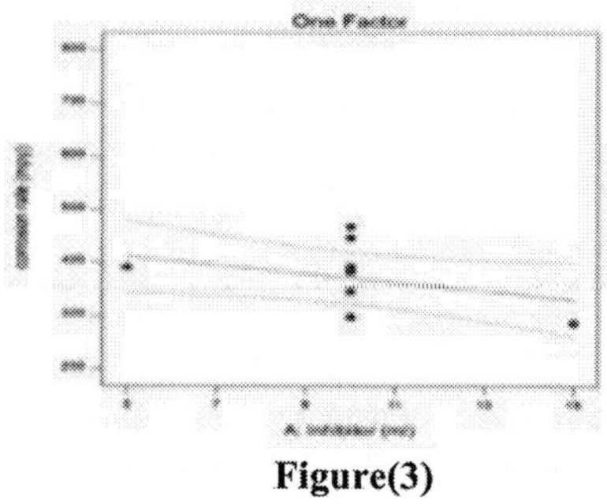

Figure(3)

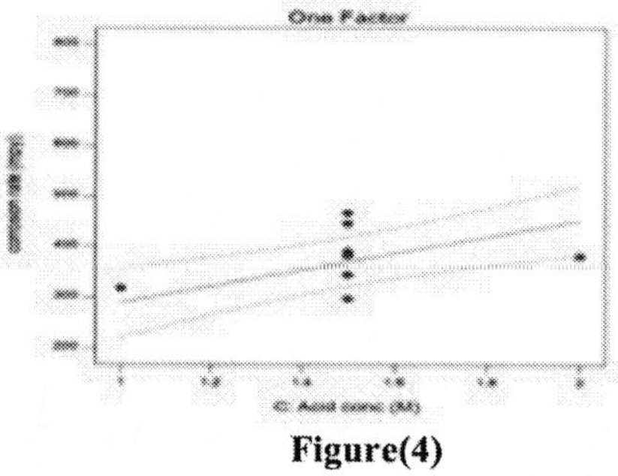

Figure(4)

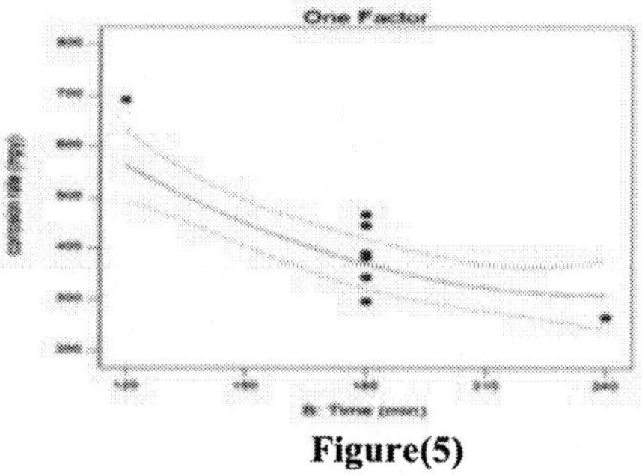

Figure(5)

The effects of two factors into the corrosion rate can be studied from the 3D plots as shown in Figure(6): 3D plot of corrosion rate Vs inhibitor concntration and acid concntration and Figure(7): 3D plot of corrosion rate Vs inhbitor concntration and immersed time.

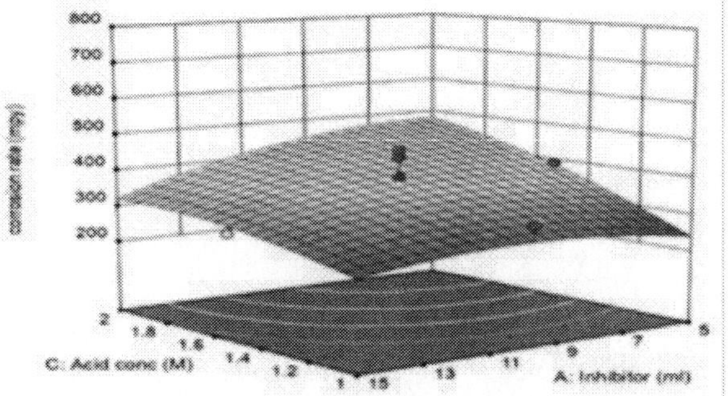

Figure(6)

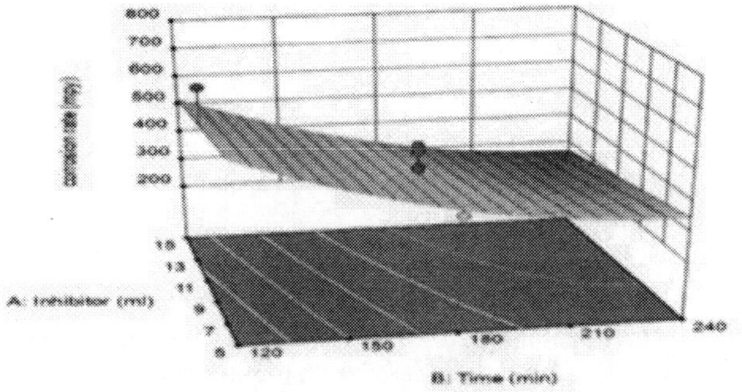

Figure(7)

The optimum conditions given by the RSM of 1M of H_2SO_4 with 5ml of berries juice extract inhibitor runs for 240 minutes gives 213.402mpy as a minimum corrosion rate but the experimental results at the same conditions gives 232.022mpy which means that they are too close as the suggested solution considered as an excellent optimum conditions.

SEM ANALYSIS

The SEM has been done for the blank mild steel sample for facilitating comparison as shown in Figure(8) which shows smooth mild steel surface compare to others. Rough surface in absence of berries juice extract is shown in Figure(9); where corrosion starting with the formation of H_2 bubbles as per the equation(3) which means that the surface of mild steel is exposed to the H_2SO_4 medium. In presence of 15 ml of berries juice extract, the surface of mild steel is protected by the adsorption of the organic compounds that are present in the berries juice extract where fewer cracks and pits can be noted and smooth surface is shown in Figure (10), which means that the mild steel surface is protected. (Raja, P. et al, 2008)

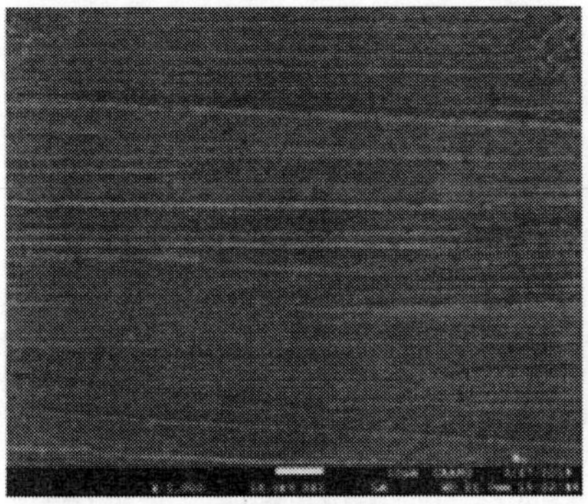

Figure(8)

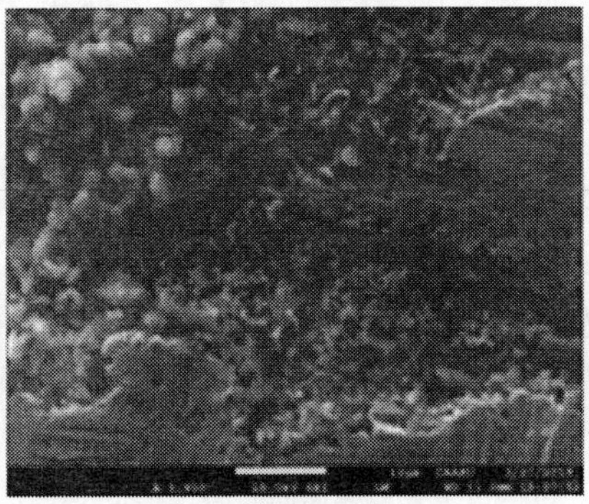

Figure(9)

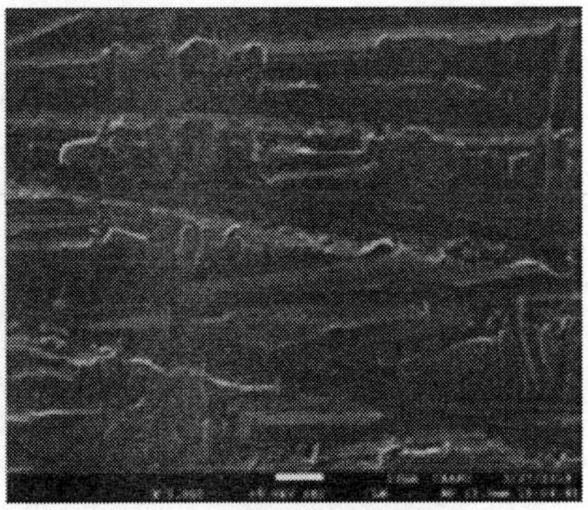

Figure(10)

EDS ANALYSIS

EDS has been used to study the chemical characterization or elemental analysis; Figure (11) shows the blank mild steel sample before being immersed in 1M H_2SO_4 in absence of berries juice extract where high Fe peak is shown, which indicates the presence of high amount of iron which reaches to 88 Wt% in the mild steel surface. Figure (12) shows the mild steel surface that has been immersed for 2 hours in 50 ml of 1M H_2SO_4 in absence of berries juice extract; the mild steel surface contains low amount of Fe (87.3 Wt%) compared to the blank mild steel sample. In Figure (13) the mild steel sample immersed for 2 hours in 50 ml of 1M H_2SO_4 in the presence of 5ml of berries juice extract, it shows high spark of Fe where the mild steel surface contains lower amount of Fe (84.8 Wt%) compared to Figure(11) and (12) which indicate lower iron dissolution rate in 1M H_2SO_4 medium containing berries juice extract (Alaneme, K. et al,2016). A high amount of oxygen and carbon are observed in presence of berries juice extract compared to the absence of berries juice extract. Si is shown in the absence and presence of berries juice extract which means that's it comes out to the surface during the reaction with sulfuric acid.

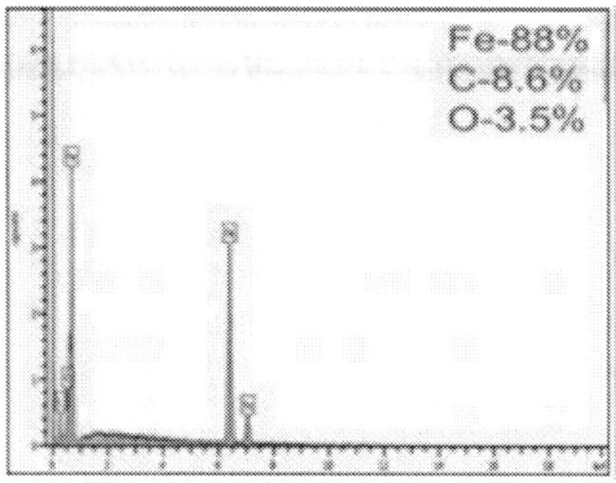

Figure(11)

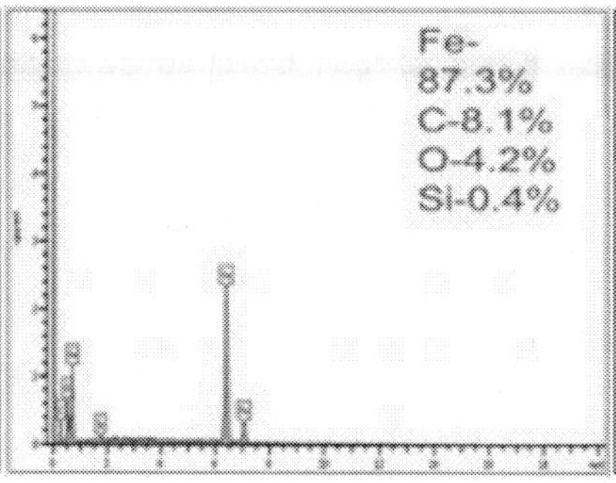

Figure(12)

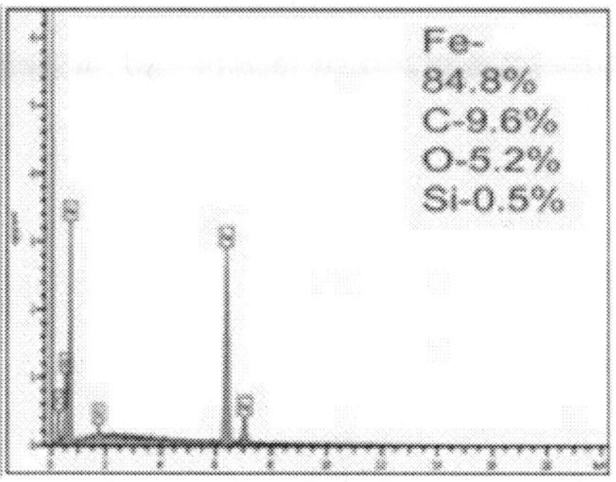

Figure(13)

GENETIC ALGORITHM OPTIMIZATION

The experimental result gives 288.367mpy as a corrosion rate. The software gives the minimum corrosion rate value of 188.866mpy from 50 runs as shown in Figure(14). A small difference can be noted as the genetic algorithm is not the exact fit with linear equations so the experimental and software results are too close to each other.

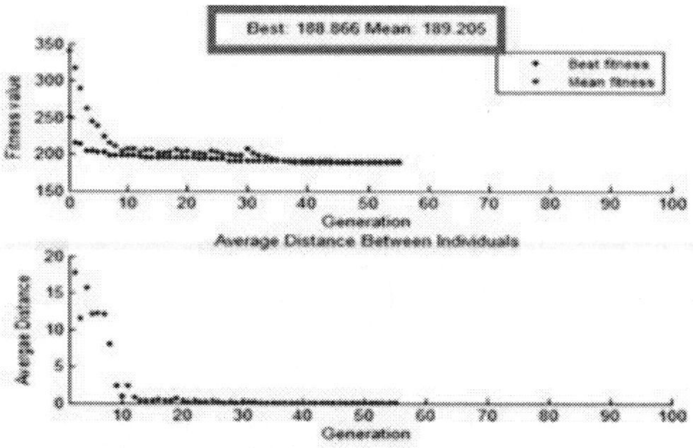

Figure(14): Genetic Algorithm optimization.

CONCLUSION

In this project, all the objectives have been achieved by designing and experimenting.
Firstly; the anticorrosive extract (berries) has been prepared, the prepared extract which contains 4 types of berries namely blueberries, blackberries, cranberries, and raspberries inhibit the mild steel corrosion in sulfuric acid medium with 41.427% efficiency in presence of 5 ml of berries extract.
Secondly, the mild steel corrosion in the presence and absence of berries juice extract were studied where the corrosion rate reaches to 1020.374mpy in absence of berries juice. By using 5 ml of berries juice extract the corrosion rate minimized to 232.022mpy.
Thirdly, the optimum conditions to inhibit the mild steel corrosion were studied by Design Expert V11. At the optimum conditions of 1M of H_2SO_4. 5ml of berries juice extract, 240 minutes and at room temperature the corrosion rate reaches to 232.022mpy as a minimum value.
The actual equation obtained from Design Expert V11 was later optimized using Genetic Algorithm. Genetic Algorithm shows that at 1.001M of H_2SO_4, 15 ml of berries juice extract and 237.208 minutes at room temperature, the corrosion rate is minimized to 188.866mpy.
Finally, mixing tank has been designed with 1.18m height and 2.5m diameter and a turbine agitator has been selected.

ACKNOWLEDGEMENT

I would like to thank god for giving me patience and power to work in my most important part of my life, my father, mother, my brother and my friends for their help and support. I would also like to express my sincere gratitude to Dr. Priy Brat Dwivedi for his support and help in developing my project idea, Mr. M.J.Varghese for his help in RSM analysis, Mr. Hatem Yazidi and Dr. Moammar Ali AlTubi for providing me the technical knowledge in Matlab optimization, Mr.Mohammed Salim Adeeb in providing me the technical help in the laboratory, Ganesh Nagraj Patil, Dr.Lakkim Setty

Rao, and Dr.Santosh Madhav Walke for their help in the calculations of design mixing tank. Last but not the least, I would like to thank National University of Science and Technology-College of Engineering for giving me this opportunity to complete the project in their laboratories and supporting me by providing me with the required chemicals.

REFERENCES

1.Alaneme,K. et al .,2016.Corrosion Inhibitory Properties of Elephant Grass(Pennisetum Purpureum) Extract Effect on Mild Steel Corrosion in 1M HCl Solution :*Alexandria Engineering Journal*.[e-journal]. 55 (8). p. 1069-1076. Available from: https://www.sciencedirect.com/science/article/pii/S1110016816300242.[Accessed: 8th May 2019].

2.Ambrish,S. et al .,2012. The Effect of Environmentally Bening Fruit Extract of Shahjan(Moringa Olifera) on the Corrosion of Mild Steel in Hydrochloric Acid Solution:*Journal of Chemical Engineering Communications*.[e-journal]. 199 (6). p. 63-77. Available from:https://www.tandfonline.com/doi/abs/10.1080/00986445.2011.570390.[Accessed: 6th May 2019].

3.Blumberg,J.,et al.,2013.Carnberries and Their Bioactive Constituents in Human Health:*Journal of Advances in Nutrition*.[e-journal]. 12 (7). p. 618-632. Available from: https://academic.oup.com/advances/article/4/6/618/4595566.[Accessed: 1th May 2019].

4.Bobinaite,R.,Viskelis,P.&Venskutonis,P.,2015.Chemical Composition of Raspberry Cultivars:*International Journal of Molecular Sciences*.[e-journal]. 16 (8). p. 18642-18663. Available from: https://www.researchgate.net/publication/306158828_Chemical_Composition_of_Raspberry_Rubus_spp_Cultivars.[Accessed: 1th May 2019].

5.Cafferty,E.,2010.*Introduction to Corrosion Science*.United States:Springer.
Roberge,P.,2012.*Handbook of Corrosion Engineering*.2nd edition.United States:McGraw-Hill company.
6.Dynamixianc,2015.*Optimal Mixer Design* .[Online]. Available from: https://www.dynamixinc.com/optimal-mixer-design. [Accessed: 14th December 2018]

7.Ibryamoval,Y. et al .,2009.Design of Mixers for Chemical and Biochemical Industry: *Trakia Journal of Sciences*.[e-journal]. 8 (6). p. 6-11. Available from: http://tru.uni-sz.bg/tsj/Vol8.Suppl.3.2010/Y.Ibryamova.pdf.[Accessed: 13th December 2018].

8.Kaume, L. et al.,2012. The Blackberry Fruit A Review on Its Composition and Chemistry, Metabolism and Bioavailability and Health Benefits:*Journal of Agricultural and Food Chemistry*.[e-journal]. 37 (7). p. 5716-5727. Available from: https://www.ncbi.nlm.nih.gov/pubmed/22082199.[Accessed: 6th May 2019].

9.Kumar.A,2012. Inhibition of Mild Steel Corrosion in HCl Solution Using Pipali(Piper Longum)Fruit Extract:*Arabian Journal for Science and Engineering*.[e-journal]. 38 (9). p. 85-97. Available from: https://link.springer.com/article/10.1007/s13369-012-0409-9.[Accessed: 9th May 2019].

10. Michalska,A.& Lysiak.G .,2015.Bioactive Compounds of Bluberries:*International Journal of Molecular Sciences*.[e-journal]. 16 (8). p. 18642-18663. Available from: https://www.ncbi.nlm.nih.gov/pmc/articles/PMC4581264/.[Accessed: 1th May 2019].

11.*Mixing Tank*,2018.[Online]. Available from: https://www.conceptdraw.com/examples/agitator-tank-symbol.[Accessed: 19th November 2018].

12. Mycourses,2016. *Response Surface Modelling-RSM*.[Online]. Available from:https://mycourses.aalto.fi/pluginfile.php/364340/mod_resource/content/1/BOS_Lecture3_2016.[Accessed: 4th May 2019].

13. Ostovari, L. et al.,2009.Corrosion Inhibition of Mild Steel in 1M HCl Soluition by henna extract:*Journal of Corrosion Science*.[e-journal]. 32 (9). p. 1935-1949. Available from: https://www.sciencedirect.com/science/article/pii/S0010938X09002157. [Accessed: 6th May 2019].

14. Petropedia,2018. *Mixing Tank*.[Online]. Available from: https://www.petropedia.com/definition/7742/mixing-tank.[Accessed: 11th December 2018].

15. Raja,P. et al.,2008.Inhibitive Effect of Black Pepper Extract on the Sulphuric Acid Corrosion of Mild Steel:*Journal of Materials Letters*.[e-journal]. 62 (9). p. 1977-2979. Available from: https://www.sciencedirect.com/science/article/abs/pii/S0167577X08001316.[Accessed: 8th May 2019].

16. Sinnot,R.&Towler,G.,2009.*Chemical Engineering Design*.5th edition.India:Elsevier.

17. Xianghong,L. et al.,2009.Inhibition Effect of 6-benzylaminopurine on the Corrosion of Cold Rolled Ssteel in H_2SO_4 Soluition:*Journal of Corrosion Science*.[e-journal]. 51 (9). p. 620-634. Available from: https://www.sciencedirect.com/science/article/pii/S0010938X09000080. [Accessed: 6th May 2019].

590434 Verification of Firefighters' Heuristics and Predictive Model Development through Big Data Analysis

Abstract:

The heuristics accumulated in the field activities of firefighters are reviewed through big data analysis of fire occurrences and emergency responses in Gyeonggi-do Province, Korea and expanded as predictive models to be utilized for proper fire prevention activities according to time, day, and target through quantitative modeling. Empirical rules with high sympathy are collected through direct interviews with firefighters. Among these, three types of high-priority rules are compared and analyzed for big data such as fire occurrence and damage history in Gyeonggi-do in 2018. Big data comparison analyses are conducted, including the number of fires and damages that occurred in Gyeonggi-do in 2018. Furthermore, fire occurrence patterns by region, day of the week, time of day, and building type are derived. In addition to using empirical rules that have been validated through research, relatively inexperienced firefighters also can make better decisions by relying on refined quantitative predictive models replacing empirical rules, including local government and time-based factors that reflect big fire occurrence data.

CONVERTING SHELTER-IN-PLACE LOCATIONS TO ZERO VULNERABILITY SAFE HAVENS

J. A. Rau
MineARC Systems America, LLC
4850 W. Ledbetter Drive
Dallas, Texas 75236
USA

A. Young
MineARC Systems, Pty Ltd
274 Welshpool Rd
Welshpool, WA 6106
Australia

M. McDermott
High Performance Building Solutions
7415 Folger Dr.
Charlotte, NC 28226
USA

CONVERTING SHELTER-IN-PLACE LOCATIONS TO ZERO VULNERABILITY SAFE HAVENS

ABSTRACT

While it is not uncommon to find shelter-in-place or safe haven locations designated across chemical facilities, what varies is the level of protection offered at these sites. Functionality can range from merely a muster point (shelter-in-place) to a positively pressurized room (safe haven). Either way, when determining the room's acceptable performance criteria, the industry generally applies ERPG-3[1] or a vulnerability factor. An acceptable vulnerability factor is often deemed the probability of a fatality being less than one percent.

In 2017, MineARC Systems and High Performance Building Solutions (HPBS) began working with a leading energy company to convert five shelter-in-place spaces to safe havens, which included additional life support systems that would ensure zero percent vulnerability. The objective of this paper is to examine the process to harden the space against contaminated air intrusion and what critical life support systems are essential for converting a shelter-in-place location to a safe haven.

INTRODUCTION

Emergency shelter-in-place (SIP) forms an integral part of any plant's broader emergency response plan. Explosions, fires, and the release of smoke or other toxic gases are some of the types of incidents that can occur despite high levels of planning and safety precautions[2].

Numerous government agencies and other organizations recommend shelter-in-place as a measure to reduce harm in the event of a chemical release. Any building may be used as a temporary measure to reduce health risk from exposure to the toxic materials by simply closing windows and doors and turning off ventilation fans and air conditioning/heating systems.

Three factors govern the effectiveness of a SIP location: 1) the tightness of the building or leakage rate, 2) the concentration of the toxic gas, and 3) the release duration. The tightness of the building is primarily based on building construction; however, temperature differential and wind speed also play a factor. Identification of the maximum toxic concentration occurs during a Facility Siting Study (FSS). FSSs are a hazard analysis that defines Maximum Credible Event (MCE) scenarios using a consequence-based approach to produce a quantitative risk assessment. Determining potential consequences is a necessary step in the process of developing a comprehensive safety plan. The FSS determines the expected release duration based on stored inventories of toxic materials. The higher the toxic concentration and longer the release duration, the less effective sheltering in place becomes.

While occupants may feel safe sheltering-in-place from the external toxic hazard, typical buildings have high leakage rates. A typical wood framed constructed building can have 3-5 air changes per hour (ACH50[3]), rendering it ineffective against most chemical releases. Additionally, sealing to an acceptable airtightness can be time consuming and expensive and even with professional air sealing it is very difficult to air seal existing structures to an acceptable tightness.

A case study for the conversion of five shelter-in-place spaces to zero percent vulnerability safe havens is presented. There are considerations for effectively sealing existing infrastructure and ensuring positive pressure to ensure contaminants do not infiltrate safe haven locations along with the critical element of life support. The critical elements of life support that are of primary importance to human survival in a safe haven are:

- Positive pressure
- Breathable air supply
- Supplemental O_2
- Removal of accumulated CO_2

---

[1] Emergency Response Planning Guidelines (ERPGs) are values developed by the AIHA Guideline Foundation's Emergency Response Planning Committee. ERPG-3 is the maximum airborne concentration below which it is believed nearly all individuals could be exposed for up to 1 hour without experiencing or developing life-threatening health effects.

2 The CSB is an independent federal agency charged with investigating industrial chemical accidents. Over the last 10 years they have completed investigations into 15 separate chemical release incidents.

[3] ACH50 is the number of times the air volume in a building changes per hour at 50 Pascal of pressure. Commercial air leakage rates are commonly reported at 75 Pascals, but this leakage study was conducted at 50 Pascals based on the targets included in the FSS report.

- Cooling and dehumidifying

CASE STUDY

In 2017 MineARC Systems and its project partner HPBS began working alongside a major energy company to convert five shelter-in-place locations into zero percent vulnerability safe havens.

The process started with the energy company completing its FSS, which included plume modeling to define the chemical release concentrations at each building location along with ideal shelter-in-place locations[4] and internal minimum positive pressure[5]. This information was used by MineARC to calculate indoor concentrations and occupant vulnerability to highlight that shelter-in-place alone would result in the vulnerability of greater than zero percent.

From there, HPBS conducted air infiltration testing following ASTM E779-10 to determine an air leakage baseline (see Table 6) for rooms that were designated to be sealed (pre-weatherization). A site assessment was then completed to review each location and determine its suitability for sealing and conversion to a safe haven. Firstly, the measured baseline infiltration rates were compared to the FEMA 453 estimates based on construction type[6] (see Figure 2). All surfaces of the room were examined for ease of conventional air sealing and mechanical equipment sealing. Rooms that included false floors, large amounts of electrical conduit, cabinetry, computer servers or windows etc. were deemed difficult to seal (i.e. control rooms). Secondly, the building or rooms open floor space was checked to ensure it could accommodate the specified number of people in an emergency. FEMA 453 section 1.6 recommends 20ft^2 per person for durations of less than 24 hours. With the FSS recommending 4 hours, it was deemed acceptable to have an unrestricted floor space of 6ft^2 per person.

Manual weatherization (air sealing) of the rooms included installing a cleanroom suspended ceiling system with gaskets and tiles that met the air permeance rate for an air barrier material, cleanroom airtight light fixtures with gaskets, sealing around electrical and plumbing, air sealing all visible air pathways to make sure the interior gypsum board was functioning as the air barrier, installing low-leakage HVAC dampers, installing an emergency shut off for the mechanical system and adding door weather-stripping and sweeps. Post-weatherization air leakage tests were conducted to determine the air infiltration rate of the SIP spaces after completion of the manual air sealing, cleanroom suspended ceiling system and HVAC work. The weatherization air sealing methods improved the air barrier integrity for all the proposed SIP spaces by between 57% and 65% (refer to Table 6).

Aerosol-based envelope sealing was used to reduce leakage rates even further. AeroBarrier® is a building envelope sealing technology that simultaneously measures and seals buildings. AeroBarrier® works by forcing an atomized sealant into orifices under pressure gradually sealing leaks. Rooms were sealed during night shift with personnel able to re-access the rooms by morning. The AeroBarrier® air sealing methods improved the air leakage for all the SIP spaces between 7% and 88% over the post manual weatherization leakage rates (refer to Table 7). AeroBarrier® is not effective for sealing large holes so manual air sealing is required for certain locations prior to using AeroBarrier®.

MineARC engineers once again calculated infiltration concentration rates along with the vulnerability rate. As the probability of a fatality was still higher than zero, MineARC then specified its AirBANK Pressure System and necessary carbon dioxide and oxygen control equipment.

SAFE HAVEN LIFE SUPPORT EQUIPMENT

Primary Life Supporting Atmosphere

A safe haven needs to provide a reliable breathable atmosphere independent of external toxic gas concentrations. Human breathing consumes oxygen (O_2) and expels carbon dioxide (CO_2), carbon monoxide (CO), water vapor, and heat. At complete rest, an average person exhales air at 17% O_2 and 3.2% CO_2. This breathing rate results in a 'resting' O_2 uptake of 0.3L/min of O_2 and a resting CO_2 discharge of 0.24L/min (Brake, Fellows, & Bates, 1999). Taking into consideration the above values, a person in a closed system such as a shelter-in-place succumbs to CO_2 poisoning well before O_2

4 Shelter-in-place location is beyond the scope of this paper, but should be determined based on the chemical hazard, facility layout, and location of personnel.

5 Positive pressure requirements are based on average wind speed at the facility. As wind speed increases so does the SIP leakage rate.

6 The FEMA 453 estimates are intended to estimate filter sizing when ASTM E779 air leakage testing is not conducted.

asphyxiation. Venter et al. (1998) found an average oxygen consumption for 12 entrapped individuals of 0.44 liters per minute, at STP, over 24 hours (including sleeping). A figure of 0.5 liters per minute over a non-sleeping entrapment period of 12 hours is a prudent, conservative design value. This corresponds to a metabolic rate of about 160W or $80W/m^2$ for a typical person with a body skin area of about $2m^2$, which is equivalent to a breathing rate of about 12.5 liters per minute of fresh air.

Consequently, life support systems are needed within a safe haven that can ensure positive pressure keeps contaminants outside, along with replacing the O_2 consumed, and removing exhaled CO_2, moisture, and heat to maintain the atmosphere within acceptable threshold limits.

Positive Pressure

Positive pressure is integral to prevent the ingress of airborne contaminants that may compromise the breathable atmosphere for occupants. Establishing a Class 1 Safe Room (FEMA 453) requires the sufficient release of filtered air from an external source, to form a positively pressurized zone.

Buildings are typically constructed to provide natural or forced ventilation to allow permanent occupancy and potentially combined with large exposed windows and surfaces which can present challenges for ensuring protection during a chemical release. When determining adequate positive pressure within a space, several structural factors need to be considered, including:

- Shelter volume
- HVAC (Heating, Ventilating, and Air Conditioning)
- Door and window tightness
- Existing opening such as vents and elevator shafts
- Indoor and outdoor temperature
- Wind speed and building orientation
- The external screening of the structure

Building design and multiple ingress points suggest the conversion of a smaller interior room within a building can provide a higher level of protection, than the whole building, as it is less subject to wind and buoyancy forces which induce the infiltration of external hazards.

Breathable air from compressed air cylinders venting directly into the SIP forms a positive internal pressure. Determining the amount of air required to reach the necessary level of positive pressure involves measuring the areas internal volume and the relationship between the internal and external atmospheric conditions.

After reaching required ACH (air changes per hour) or tightness, a positive-pressure management system must be sized for the expected shelter duration.

Breathable Air Supply

If the shelter-in-place is airtight, a supply of breathable air is necessary to support occupants over the required shelter duration. Unventilated SIPs that are tightly sealed cannot be occupied for long periods without the risk of high carbon dioxide levels. A secondary source of breathable air must be circulated within the shelter to meet occupancy over the required amount of time.

Clean, breathable air is supplied to a chemical, biological and radiological (CBR) safe room in a few ways: by trapping air inside the room and minimizing the air exchange, passing contaminated air through a filter (not practical for high concentration chemical releases or for chemicals that cannot be filtered) to purify it as it is supplied to the room, and the supplemental supply of oxygen through high pressure compressed air cylinders.

The air supply must contain enough oxygen in its mix, and not contain contaminants like carbon monoxide, carbon dioxide, water, or any other toxins that poison safe haven occupants during entrapment.

Table 1 lists international standards for flowrates for breathable compressed air for self-contained breathing apparatus (SCBA), supplied air respirators (SAR) and refuge chambers.

Table 1 – Standards for acceptable air supply requirements

Air supply requirements	Relevant Standard
Open-circuit positive-pressure SCBA, including all components and accessories except the air cylinder (shell) must resist the permeation and penetration of HD and GB chemical agents when tested for a period of 6 hours on an upper-torso manikin connected to a breathing machine operating at an airflow rate of 40 liters per minute (lpm), 36 respirations per minute, 1.1 liters tidal volume.	NIOSH 42 CFR Part 84 Respiratory Protective Devices
The flow rate should be at least 85L/min (3cfm) per person to remove expired CO2 and CO but at least 255L/min (9cfm) to provide adequate cooling and maintain comfortable conditions. The air inflow is assumed to be at a temperature of 26.5°C and 65% relative humidity	ITA Report No. 14 – Guideline for the provision of refuge chambers in tunnels under construction, May 2014
For manually operated blower type air-hose respirators, the blower shall be capable of delivering air to the space enclosed by the facepiece at a flow rate of at least 85 L/min.	AS/NZS 17:16:2003 Respiratory Protective Devices

The NIOSH standard of 40L/min is for SCBA apparatus (i.e. mouthpiece or face mask) where dilution of CO_2 is not a concern. Conversely, the ITA and Australian standards specify 85L/min and are concerned with both oxygen consumption and carbon dioxide dilution. It is important to note that three times the amount of air is required to maintain comfortable conditions inside a refuge chamber, when compared to simply breathing (Health & Safety Executive UK, 2006).

Unless the positive pressure system is supplying breathable air at a rate that maintains oxygen and compressed gas levels within acceptable ranges (dependent on safe haven occupancy and duration), a CO_2 scrubber and supplementary oxygen supply are required.

Supplemental O_2

O_2 is an essential requirement for cellular activity, and therefore, the absence of O_2 for relatively short periods results in death. Humans within an enclosed environment consume O_2 and if not replaced, would result in an irrespirable atmosphere. An O_2 concentration of less than 11% poses a high risk of death.

Retaining the oxygen levels between acceptable limits is critical. Various international standards outline acceptable O_2 limits for a breathable environment; examples are shown in Table 2.

Table 2 - Standards for allowable O_2 levels

Minimum O_2 Content (% by volume)	Maximum O_2 Content (% by volume)	Relevant Standard
18.5%	23.0%	Mines Safety and Health Administration 30 CFR Parts 7 and 75 Refuge Alternatives for Underground Coal Mines; Final Rule
19.5%	23.5%	Occupational Safety and Health Administration 29 CFR 1910.146 Permit-Required Confined Spaces for General Industry
19.5%	23.5%	Australian Standard AS 2865–2009 Safe working in a confined space

OSHA and Australian Standards for working in a confined space specify an acceptable lower limit of 19.5%, which is consistent with many other standards around the world. The slightly lower MSHA limit of 18.5% is based on a person at rest, not working, therefore, not likely to have trouble in breathing; conditions that are likely to be present in a refuge chamber.

O_2 itself is not flammable but supports and accelerates combustion. OSHA defines O_2 enriched mixtures or atmospheres as those containing more than 23.5% O_2 by volume (OSHA, 2013). In O_2-enriched atmospheres, the reactivity of O_2 significantly increases the risk of ignition and fire.

Within a safe haven, pure O_2 must be utilized in conjunction with a CO_2 removal system.

Removal of Accumulated CO_2

Protection within a safe haven is substantially increased by filtering CO_2 from the air. CO_2 is being produced continuously by the cells of our body and is a byproduct of respiration. The inability to expel this gas or an environment with high concentrations of CO_2 leads to CO_2 poisoning. Breathing air with more than 3% CO_2 can induce dizziness, headaches, and vision disturbance. Prolonged exposure at greater than 6% could result in unconsciousness and death. Listed in Table 3 are examples of international standards of satisfactory CO_2 removal requirements for a breathable environment.

Table 3 – Acceptable CO_2 removal requirements

CO_2 Removal Requirements	**Relevant Standard**
CO_2 removal components shall remove CO_2 at a rate of 1.08 ft^3 (30.5 L)/h/person.	Mines Safety and Health Administration 30 CFR Parts 7 and 75 Refuge Alternatives for Underground Coal Mines; Final Rule
OSHA has established a Permissible Exposure Limit (PEL) for CO2 of 5,000 parts per million (ppm) (0.5% CO2 in air) averaged over an 8-hour workday (time-weighted average or TWA:)	Occupational Safety and Health Administration
5.2.8 Atmospheric monitoring. Atmospheric monitoring There should be the capability to monitor temperature continuously, and for carbon monoxide, carbon dioxide, and oxygen in the chamber during use in both externally supported or standalone modes. The concentration of carbon dioxide should not exceed 10000 ppm.	ITA Report No. 14 – Guideline for the provision of refuge chambers in tunnels under construction, May 2014

The CO_2 that is exhaled by the occupants within the safe haven must be removed by, constant flushing by an external breathable air source (e.g. compressed air cylinders used for positive pressure) or by the regeneration of the internal atmosphere with a CO_2 absorbent (scrubber).

Different regulations around the world, state varied human production rates between 30–50L per hour per person. When considering an acceptable production rate; it must be noted that the rate of breathing is primarily triggered by the CO_2 content of inspired air with higher CO_2 levels triggering faster respiration rates (Brake, Fellows, & Bates, 1999) (i.e. if the maximum allowable CO_2 concentration inside the safe haven is 1%, a lower production rate is acceptable compared to a 3% CO_2 limit).

If CO_2 removal mechanisms are inadequate, the occupants can develop CO_2 toxicity by re-breathing their own exhaled CO_2.

Cooling and Dehumidifying

Thermal tolerability with a safe haven should be addressed during the design phase. Assumptions must be made in preparation for safe haven design in regard to the number of personnel, activity level, available hydration and heat production. The minimum airflow to maintain a comfortable environment is set by taking these assumptions into account as well as the input temperature and humidity of supplied air.

The healthy human body maintains its internal temperature around 37°C. An essential requirement for continued normal body function is that the deep body temperature is maintained within the narrow 1°C limit (Epstein, Moran, 2006). A change of body temperature exceeding 1°C occurs only during illness or when environmental conditions surpass the body's ability to cope with extreme temperatures.

When a person is subject to extreme heat and humidity over some time, the person may experience a very rapid progression of symptoms or exhibit a few of the less significant symptoms before falling victim to more severe forms of heat-related illness.

The acceptable cooling requirements for underground refuge chambers are defined in various international standards, examples of which are listed in the below table.

Table 4 – Standards for acceptable cooling requirements

Cooling Requirements	Relevant Standard
Simulated emergencies, in which a full complement of people has occupied a refuge chamber for a significant period, indicate that humidity and temperature can increase very rapidly to potentially heatstroke-inducing levels. Refrigerated air conditioning is strongly recommended for both externally supported and stand-alone refuge chambers to counter this potentially dangerous problem. Inevitably, this places a heavy demand on the stand-alone power supply, but there are systems available that can cope, and at an acceptable cost.	Department of Mines, Industry Regulations and Safety, 2005 Refuge Chambers in Underground metalliferous mines – guideline: Safety and Health Division, Department of Industry and Resources Western Australia, 32 pp
Apparent temperature in the fully occupied refuge chamber must not exceed 95 °F.	Mines Safety and Health Administration 30 CFR Parts 7 and 75 Refuge Alternatives for Underground Coal Mines; Final Rule
It should be possible to maintain an internal temperature not exceeding 30 °C wet bulb globe temperature for at least 24 hours	ITA Report No. 14 – Guideline for the provision of refuge chambers in tunnels under construction, May 2014

MSHA regulations specify a maximum internal apparent temperature[7] of 95°F (35°C). NIOSH specifies an equivalent maximum wet bulb temperature of 28.9°C (NIOSH, 2007).

The temperature maintained inside the safe haven is determined by the metabolic heat liberated by the occupants, dissipated to the external environment by the heat transfer characteristics of the room or building. The external temperature primarily influences the heat transfer while the humidity is the combination of moisture produced by the occupants (via exhaled air and sweat) and water liberated by CO_2 scrubbing chemicals.

It is important to remember that in most cases, before sheltering-in-place, the air conditioning system will need to be shut down (unless it is a split system). For short durations, artificial cooling is not a critical requirement unless external temperatures are extraordinarily high, and/or the safe haven's volume is small compared to its occupancy.

INFILTRATION TESTING

Background

High Performance Building Solutions Inc. (HPBS) was contracted to perform an air infiltration study for spaces identified as candidates to be used as shelter-in-place (SIP) in the event of an accidental chemical release. This SIP strategy uses the existing structure and its indoor atmosphere to separate individuals from a hazardous outdoor atmosphere temporarily.

A series of pre-weatherization air leakage tests were conducted to determine the baseline air infiltration rate of potential SIP spaces. Manual air sealing, a cleanroom suspended ceiling system and low-leakage HVAC dampers were installed to reduce the overall air infiltration of the SIP spaces and to close off air leakage pathways from non-protected spaces in the building through the HVAC system. Post-weatherization air leakage tests were conducted to determine the air infiltration rate of the SIP space after the completion of weatherization and HVAC work. The SIP spaces were additionally air sealed using the proprietary AeroBarrier® system to further reduce infiltration rates.

Scope

Air sealing work was completed, and low-leakage HVAC dampers were installed to reduce the overall air infiltration of the SIP spaces and to isolate the HVAC from non-protected spaces in the building. Post-weatherization air leakage tests were conducted to determine the air infiltration rate of the SIP space after the completion of air sealing, suspended ceiling and HVAC work. AeroBarrier® air sealing was conducted on SIP spaces as a pilot project on September 18 through September 21, 2017. Post AeroBarrier® air leakage tests were conducted on September 21 and September 22, 2017, to determine the air infiltration rate of the SIP space after the completion of the AeroBarrier® sealing.

7 Apparent temperature is the temperature equivalent perceived by humans, caused by the combined effects of air temperature and relative humidity.

Following ASTM E 779-10, pressurization air leakage tests were conducted on the following spaces recommended in the Facility Siting Study (FSS).

Table 5: SIP specifications

SIP Space	Occupancy	Floor Area (Square Feet)	Volume (Cubic Feet)
SIP Space 1	40	990	10,481
SIP Space 2	6	955	8,210
SIP Space 3	5	391	3,917
SIP Space 4	10	1,697	16,968
SIP Space 5	40	293	2,637

Air Leakage Test Results

Positive pressure air leakage test results are reported in four different ways:

CFM4 is the quantitative volume of air passing through an enclosure under a pressure difference of 4 Pascals (equivalent to a 5mph wind). This is a measurable physical quantity and the value recommended in the FSS to eliminate toxic gas ingress into the SIP spaces.

CFM25 is the quantitative volume of air passing through an enclosure under a pressure difference of 25 Pascals (0.1-in w.g.). FEMA 453 requires Class 1 Safe Havens be pressurized to 25 Pascals. This is a measurable physical property.

CFM50 is the quantitative volume of air passing through an enclosure under a pressure difference of 50 Pascals (0.2-in w.g.).

ACH50 is the quantitative volume of air passing through an enclosure under a pressure difference of 50 Pascals (0.2-in w.g.) normalized by the enclosure volume. ACH is the abbreviation for air changes per hour. It is the number of times the air volume in a building changes per hour at 50 Pascals of pressure.

Table 6 below shows the CFM50 and ACH50 for each of the SIP spaces pre-weatherization and post-weatherization along with the percentage reduction in leakage. SIP Space 1 and SIP Space 5 did not have pre-weatherization results as initially the whole building was tested. Smaller spaces were selected as the whole building was too large to feasibly convert to a SIP space (re-testing of the new SIP spaces was not completed). The control room, SIP space 4, was abandoned in favor of a smaller adjacent room due to high leakage post manual weatherization and the computers that could not be shut down, which prevented the use of AeroBarrier® in the space.

Table 6: Post-weatherization leakage results

Safe Haven Space	CFM50		ACH50		% Reduction
	Pre-Weatherization	Post-Weatherization	Pre-Weatherization	Post-Weatherization	
SIP Space 1	N/A	1,871	N/A	10.7	N/A
SIP Space 2	3,623	1,560	26	11.4	57%
SIP Space 3	385	135	6	2.1	65%
SIP Space 4	N/A	N/A	N/A	N/A	N/A
SIP Space 5	N/A	655	N/A	19.4	N/A

The AeroBarrier® air sealing was prescribed a target. According to the AeroBarrier® simultaneous air leakage calculations, all the spaces met and surpassed this airtightness target; AeroBarrier® sealing was halted at this point. Table 7 shows the CFM50 and ACH50 for each of the SIP spaces as tested by HPBS post AeroBarrier® along with the percentage reduction. The HPBS results were utilized to calculate the amount of positive pressure required to convert the rooms to positive pressure safe havens.

Table 7: Post AeroBarrier® leakage results

Safe Haven Space	CFM50		ACH50		% Reduction
	Post-Weatherization	Post AeroBarrier®	Post-Weatherization	Post AeroBarrier®	
SIP Space 1	1,871	455	10.7	2.6	76%
SIP Space 2	1,560	182	11.4	1.3	88%
SIP Space 3	135	125	2.1	1.9	7%
SIP Space 4	N/A	173	N/A	3.3	N/A
SIP Space 5	655	138	19.4	3.1	79%

The AeroBarrier® air sealing methods improved the air barrier integrity for all the proposed SIP spaces between 7% and 88% over the post manual weatherization leakage rates. Air leakage was still observed behind many electrical penetrations including HVAC thermostats, and fire and smoke alarms. These were covered with plastic to protect them from the AeroBarrier® Sealant but had hidden holes behind them.

Figure 1 displays the post AeroBarrier® leakage rate in comparison to the FEMA 453 estimates for air leakage rates based on construction types. FEMA 453 estimates are reported in square feet of floor area at 25 Pascals and do not take into consideration the ceiling height and volume of the space.

Figure 1: Post AeroBarrier® results compared to FEMA 453 estimates

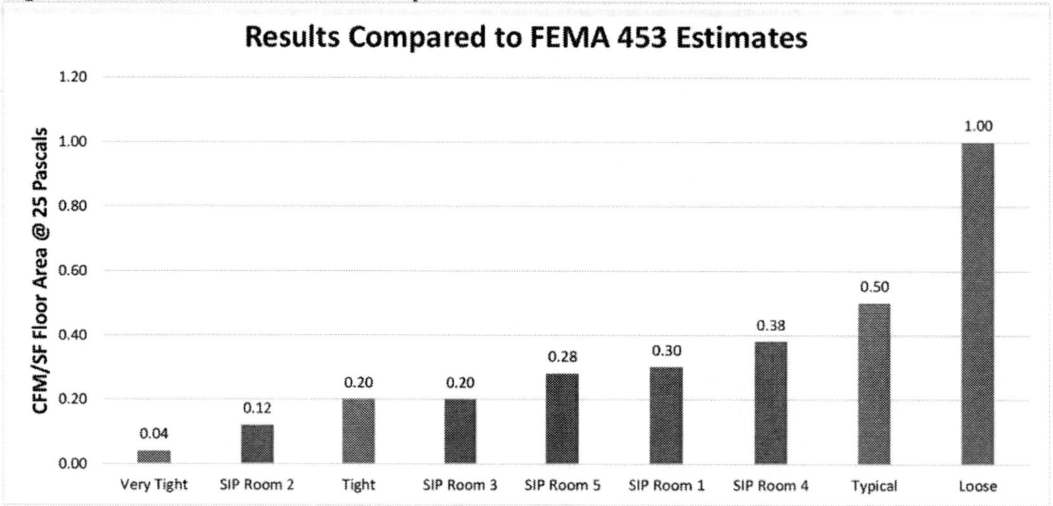

Figure 2 is an estimated leakage rate based on construction type to estimate the filtration and positive pressure system size under the FEMA 453 standard. All the SIP locations were between tight and typical except for SIP Room 2, which was a small prefabricated metal constructed portable building.

Figure 2: Estimated leakage rate per square foot of floor space to achieve 0.1 iwg (25Pa) overpressure

Construction Type	cfm per square foot of floor area
Very tight: 26-inch thick concrete walls and roof with no windows	0.04
Tight: 12-inch thick concrete or block walls and roof with tight windows and multiple, sealed penetrations	0.20
Typical: 12-inch thick concrete or block walls with gypsum wall board ceilings or composition roof and multiple, sealed penetrations	0.50
Loose: Wood-frame construction without special sealing measures	1.00

Table 8 shows the combined AeroBarrier®, and manual air sealing methods improved the airtightness for all the proposed SIP spaces between 68% and 95% over the original leakage rates.

Table 8: Air infiltration improvement for SIP spaces

Safe Haven Space	ACH50			% Reduction
	Pre-Weatherization	Post-Weatherization	Post AeroBarrier®	
SIP Space 1	N/A	10.7	2.6	76%
SIP Space 2	26.5	11.4	1.3	95%
SIP Space 3	5.9	2.1	1.9	68%
SIP Space 4	N/A	N/A	3.3	N/A
SIP Space 5	N/A	19.4	3.1	84%

Figure 3 compares the ACH50 pre-weatherization, post-weatherization and AeroBarrier® air infiltration results.

Figure 3: Pre-weatherization, post-weatherization and AeroBarrier® comparison

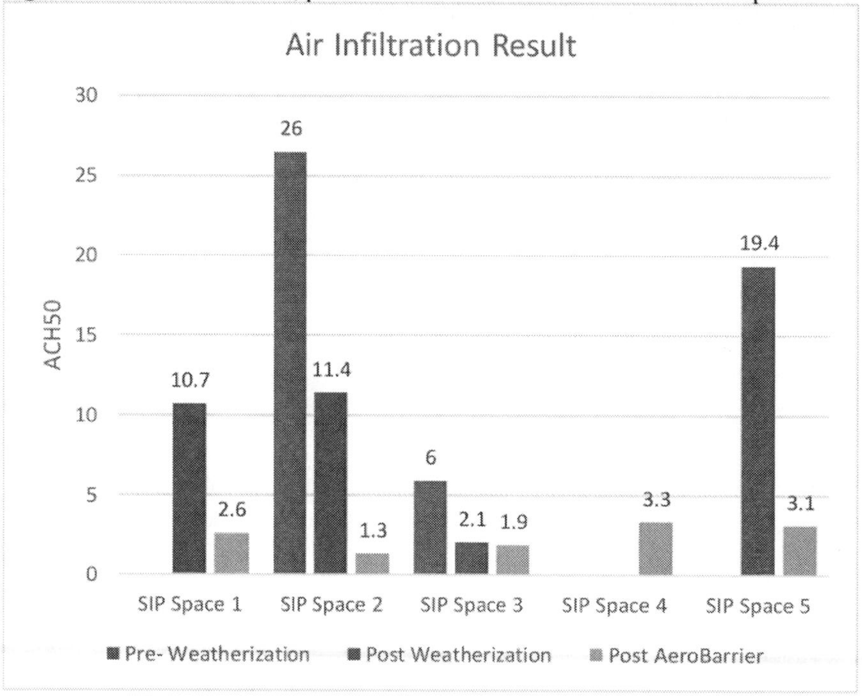

Life Support Systems

MineARC re-calculated the vulnerability for each SIP space based on the post AeroBarrier® leakage rates. All SIP spaces still had a vulnerability of 100% and therefore required the installation of independent life support systems to be deemed zero percent vulnerability safe havens.

Figure 4 shows the toxic gas concentration, duration of the release, and vulnerability for each SIP space. As the largest MCE was a full tank rupture the toxic concentrations were extremely high, but the duration was short at one minute.

Figure 4: SIP space vulnerability post AeroBarrier®

Safe Haven Location	TOXIC GAS				
	Worst Case Threat Scenario	Toxic External [ppm]	Duration of Reslease [min]	ACM	Vulnerability
SIP Space 1	Tank Rupture	585,000	1	0.032	100%
SIP Space 2	Tank Rupture	652,000	1	0.042	100%
SIP Space 3	Tank Rupture	660,000	1	0.007	100%
SIP Space 4	Tank Rupture	412,000	1	0.042	100%
SIP Space 5	Tank Rupture	600,000	1	0.028	100%

The leakage rate at 4 Pascals of pressure difference is slightly more than would occur under a 5 mile per hour wind used in the dispersion model in the FSS report based on climatic data for the facility (to maintain positive pressure and neutralize air infiltration). Post-weatherization infiltration rates in cubic feet per minute at 4 Pascals of pressure for each of the SIP spaces is shown in Table 9.

Table 9: Post AeroBarrier® air leakage results at CFM4

Safe Haven Space	Volume (Cubic Feet)	CFM4
SIP Space 1	10,481	99
SIP Space 2	8,210	32
SIP Space 3	3,917	24
SIP Space 4	3,110	45
SIP Space 5	2,637	21

The air leakage test results provide a mathematical formula that can be used to calculate the amount of breathable compressed air cylinders (437ft^3) that must be added to each candidate SIP space to maintain a 4 Pascal positive pressure to overcome a 5mph design wind coincident with a toxic release.

MineARC Systems was contracted to install life support systems to convert the SIP spaces to positive pressure safe havens. The MineARC AirBANK Positive Pressure System provides rapid pressurization, which is activated and maintained using the AirBANK Control via a simple human-machine interface (HMI) touch screen. MineARC's integrated Aura-FX Gas Monitor ensures breathable air automatically remains within acceptable limits. Additionally, the installation of supplementary oxygen and carbon dioxide scrubbing systems in safe havens, are as needed.

Figure 5 shows the number of 437 cubic feet cylinders required for each safe haven location. The minimum breathable air per person is 2.5cfm (70L/min) per person based on MineARC's experience and testing within the refuge chamber industry.

Figure 5: Quantity of cylinders required for 4Pa positive pressure

	SAFE HAVEN SPECIFICATIONS		POSITIVE PRESSURE (@ 4Pa)						
Safe Haven Location	Occupancy	Duration	CFM	CFM per Floor Space	ACH	No. Cylinders	No. Cylinder Racks	Breathable Air per Person [CFM]	Breathable Air per Person [LPM]
Safe Haven Space 1	40	4	99	0.10	0.6	55	5	2.5	70.8
Safe Haven Space 2	6	4	32	0.03	0.2	18	2	5.4	152.9
Safe Haven Space 3	5	4	24	0.06	0.4	14	2	4.8	135.9
Safe Haven Space 4	10	4	45	0.14	0.9	25	3	4.5	127.4
Safe Haven Space 5	40	4	21	0.07	0.5	12	2	0.6	17.0

The maximum CO_2 concentration is at 1% and O_2 between 19.5% and 23.5% based on OSHA limits (refer to Table 2). Due to the volume and number of people in Safe Haven Space 5, it required supplemental oxygen and carbon dioxide scrubbing to maintain the internal gas concentrations within specified limits. Three sodium chlorate oxygen candles and two AirGEN CO_2 scrubbers were included in the design.

Figure 6: Internal gas concentrations in safe havens

	SAFE HAVEN SPECIFICATIONS		POSITIVE PRESSURE (@ 4Pa)			02 & CO2 CONCENTRATIONS (@ 4Pa)			
Safe Haven Location	Occupancy	Duration	CFM	CFM per Floor Space	ACH	Indoor O_2 %	No. O_2 Candles	Indoor CO_2 %	No. AirGEN Scrubbers
Safe Haven Space 1	40	4	99	0.10	0.6	20.45	0	0.50	0
Safe Haven Space 2	6	4	32	0.03	0.2	20.84	0	0.17	0
Safe Haven Space 3	5	4	24	0.06	0.4	20.77	0	0.23	0
Safe Haven Space 4	10	4	45	0.14	0.9	20.66	0	0.31	0
Safe Haven Space 5	40	4	21	0.07	0.5	18.67	3	2.04	2

In 2018 MineARC completed installation and training on the five safe haven spaces. Each safe haven was tested during the commissioning process to ensure it could maintain acceptable positive pressure.

CONCLUSIONS

The conversion of shelter-in-place (SIP) locations to zero percent vulnerability safe havens is a four-step process consisting of the following:

1) Identifying the correct SIP location based on ease of sealing and small volume based on the required occupancy.
2) Manual air sealing of SIP spaces.
3) AeroBarrier® sealing of SIP spaces.
4) Installation of life support equipment.

Identifying SIP spaces that can seal from the outset is crucial to a successful result. In the case study, three of the five original locations were abandoned in favor of rooms in adjacent locations due to high leakage rates or reducing the size to make the project more cost effective. Control rooms pose the most difficulty in trying to seal due to the false floor and a myriad of cable tray and pass-throughs installed for the computer systems. The smaller the room and the more robust the construction (i.e. prefabricated metal or concrete buildings), the better it seals.

Manual air sealing needs to be overseen by someone with familiarity in building tightness. While most of the tasks are general construction, standard contractors are not trained to work at the level of precision needed to ensure proper sealing. Even a tiny orifice or ceiling tile not sealed correctly can ruin a leakage test result. HPBS and MineARC Systems have developed a system for ensuring airtightness in construction methodology and by avoiding un-sealable spaces or spaces that are not cost effective to seal.

For future construction projects, incorporation of SIP spaces should occur during the design phase. Air barrier commissioning is standard in new construction and should include the following tasks at a minimum: design review of the plans with specifications for the air barrier and mechanical system - specifically as these relate to spaces designated for use as safe havens; quality assurance inspections during construction to ensure a continuous air barrier is in place; and final air leakage testing of the continuous air barrier to ensure the leakage rate of the safe haven spaces are within acceptable limits.

Complementary life support equipment is essential to sustain life within an airtight environment. For personnel to remain within the sealed safe haven, the environment must be habitable. Air is to be free of toxins such as carbon dioxide and acceptable levels of oxygen maintained. Positive pressure systems reduce the ingress of toxins and provide some or all of the required source of breathable air. Additional equipment such as compressed air, oxygen cylinders and air scrubbers are used to meet the shortfall. Furthermore, internal conditions must consider the human condition, supporting workable body temperature and humidity.

Reducing the risk of harm to personnel is a priority for emergency response planning. While the likelihood of an MCE occurring may be low, the risk to on-site personnel in the event of an incident is high. Converting shelter-in-place (SIP) locations to zero vulnerability safe havens improves emergency response.

REFERENCES

ASTM E779-10, *Standard Test Method for Determining Air Leakage Rate by Fan Pressurization*, ASTM International, West Conshohocken, PA, 2019, www.astm.org

Brake, Fellow, & Bates (1999). *Criteria for the design of emergency refuge stations for an underground metal mine*. Aus. IMM Proceedings, No.2, 1999, pp 1–7.

Department of Mines and Petroleum, 2013, *Refuge chambers in underground mines — guideline: Resources Safety*, Department of Mines and Petroleum, Western Australia, 41 pp.

Epstein, Y., & Moran D. (2006). *Thermal Comfort and the Heat Stress Indices,* Industrial Health, 44, 388–398.

FEMA 453. (2006). *Design Guidance for Shelters and Safe Rooms.* Federal Emergency Management Agency

Health & Safety Executive UK (2008). *Guidance and Information on the role and design of safe havens in arrangements for escape from mines.* Retrieved from http://www.hse.gov.uk/pubns/mines08.pdf

ITA Working Group 5 (2016). *Provisions for refuge chambers in tunnel projects under construction.* ITA Report N°014 - V2 / March 2018

Mines Safety and Health Administration (2008). *30 CFR Parts 7 and 75 Refuge Alternatives for Underground Coal Mines; Final Rule.*

National Institute for Occupational Safety and Health (2007). *Research Report on Refuge Alternatives for Underground Coal Mines.* Retrieved from http://www.cdc.gov/niosh/docket/archive/pdfs/NIOSH-125/125 ResearchReportonRefugeAlternatives.pdf

Occupational Safety & Health Administration (2013). *Permit-required confined spaces.* Retrieved from http://www.osha.gov/pls/oshaweb/owadisp.show_document?p_table=STANDARDS&p_id=9797

Suming Du. (2014). *On the effectiveness of shelter-in-place as a measure to reduce harm from atmospheric releases. Journal of Emergency Management, Vol. 12, No, 3, May/June 2014.*

Venter. J. (1998). *Portable Refuge Chambers: Air or Tomb in Underground Escape Strategies.* School for Mechanical and Materials Engineering, Potchefstoom University, Republic of South Africa

Worksafe Australia National Standard (2005). *Australian Standard AS: NZS 1716:2003 Respiratory protective devices.*

Yantosik. G. (2006). *Shelter-in-Place Protective Action Guide.* The Chemical Stockpile Emergency Preparedness Program (CSEPP). U.S. Department of Energy

589008 How the Field of Human Factors Helps Solve the Riddle of Serious Injuries and Fatalities

Abstract:

Process and occupational safety are often treated as distinct endeavors, with the former tending to focus on facility system function and the latter on individual worker wellbeing. Though other differences certainly exist between the two, both endeavors are generally purposed to prevent incidents that can cost lives, health, time, and money, in addition to employee morale and public trust. The overarching thesis of this paper is that the field of Human Factors is uniquely positioned to effect positive change in both of these major safety categories.

The backdrop for our argument is the recent finding that while injury and illness incidence rates published by the Bureau of Labor statistics have decreased in recent years, the occurrences of serious injuries and fatalities have declined at a significantly slower rate. We discuss possible explanations for this finding but note that regardless of the preferred explanation, the trend should motivate scientists, engineers, and other safety professionals to revisit long-held accident causation theories holding that near misses – these are incidents that could have resulted in serious injuries and fatalities but did not – are precursors of future serious incidents. While this idea still has merit, the just-mentioned statistics warrant reconsideration of the relations between serious and less-serious incidents to explain why the rates of the two are not decreasing in lockstep. We suggest these data indicate additional (human) factors are involved.

To address this riddle and to support our general thesis, we (a) review core tenets of process safety, occupational safety, and the field of Human Factors, (b) point out that the human is always an element common to both process and occupational safety contexts, and (c) discuss how preventing or mitigating human error can benefit both. We offer specific examples from both safety contexts to show how Human Factors perspectives and techniques can help. Examples touch on standards, policies, and other administrative controls (e.g., procedures), human-machine interfaces of both workplace hardware (e.g., valves) and software (e.g., SCADA systems), environments (e.g., physical, regulatory, and cultural), and communications between personnel (e.g., elements of team resource management). We conclude that fundamental Human Factors considerations can help prevent and mitigate human performance gaps and therefore increase the rate of reduction in serious injuries and fatalities in both process and occupational safety.

589233 Initial Conditions Used in Atmospheric Dispersion Models:Issues with Commonly Used Assumptions for Gas Jets and Plumes

Abstract:

When simulating jets and plumes, all atmospheric dispersion models require initial conditions that appropriately parameterize the behavior of a gas release to the atmosphere. Often, process conditions are such that loss of primary containment of a gas will result in choked flow conditions. A great deal of experimental and theoretical consideration has been given to the flow from stagnation conditions to choked flow conditions, but less experimental work has centered on describing the flow through the depressurization process to atmospheric pressure where atmospheric dispersion models would be applicable. Traditional modeling of the depressurization process has taken the simplified approach of assuming a (depressurized) "top hat" profile (laterally constant) with application of the principles of conservation of mass, momentum, and energy. Such assumptions are known to result in predicted depressurized velocities that exceed the (sonic) velocity at the point of loss of containment, and because of the energy balance, the associated temperatures are significantly less than temperatures at the choked conditions. Physically, these effects could possibly self-cancel because as jet velocity slows, the decrease in kinetic energy of the jet will increase the jet temperature. However, no atmospheric dispersion model used for routine hazard assessment purposes accounts for kinetic energy in the energy balance. The traditional modeling approaches of the depressurization process are reconsidered based on analysis of available data, and an alternative approach to modeling the depressurization process is proposed. Example cases showing the important differences between the two approaches are discussed.

Fault Tree Uncertainty Analysis

Raymond "Randy" Freeman
S&PP Consulting
12303 Lake Shore Ridge
Houston, TX 77041
Cell: 713 408 0357
Rafree@yahoo.com

For Presentation at the 2020 Spring National AIChE Meeting
Houston, Texas

March 30, 2020

ABSTRACT

Fault tree analysis (FTA) is a widely used methodology in the process industries. FTA is used for the development of failure mechanisms, computation of failure frequencies and the determination of the probability of failure on demand of safety systems. Much of the data used in a FTA study are uncertain. For example, the failure rate of a pump is often not known with great precision. Likewise the failure rates of instrumentation are often known only within some defined limits. The common practice, used by analysts in the quantification of a fault tree, is to use the most likely or best guess as to the needed failure rate data. The use of best guess values as data inputs to the quantification of a fault tree creates uncertainty in the computed results.

This paper presents a general methodology for the determination of the impact of uncertainty on the results of a fault tree study. The general methodology is based on the mathematics of propagation of error and variance contribution analysis. For a fault tree that contains 10 event minimal cut sets, the methodology will compute the variance in the probability (or frequency) of the top event with a maximum error of 5%. For 5 event minimal cut sets the error in variance of the top event probability (or frequency) will be no more than 2%. An example is presented to illustrate the application of the fault tree uncertainty analysis methodology to a real world problem.

Keywords

Quantitative Risk Assessment, Fault Tree Analysis, Uncertainty Analysis

1. Background

Fault tree analysis is a widely used method for representation of failure mechanisms in chemical plants. The methodology is described in the CCPS CPQRA guideline book [Ref. 1] and is extensively discussed in Chapter 9 of Lees [Ref. 2]. Lacking in these discussions is a systematic method to quantify the uncertainty of the results of the fault tree calculations.

Fault tree analysis has also been used in the aircraft and aerospace industries, nuclear power plant review and is recognized by the OSHA Process Safety Management (PSM) standard [Ref. 3] as a method for process hazards analysis.

2. Review of Error Propagation and Variance Contribution Analysis (VCA) Methodology

The mean and variance of a function of random variables can be approximated using the method described by Haugen [Ref. 4] and applied by Freeman [Ref. 5, 6, 7]. Define an arbitrary function of a set of random variables, x_i, as:

Let

$$Y = F(x_i) \qquad \text{(Eq 1)}$$

The mean of Y can be estimated using the following approximation:

$$E(Y) = F[\,E(x_i)] \qquad \text{(Eq 2)}$$

Where:

$E(Y)$ = expected value of random variable Y = mean of Y

$E(x_i)$ = expected value of random variable x_i = mean of x_i

The variance of Y can likewise be estimated as:

$$V(Y) = \sum_{i=1}^{n} \left[\frac{\partial Y}{\partial x_i} \right]^2 V(x_i) \qquad \text{(Eq 3)}$$

Where:

$V(Y)$ = variance of random variable Y as defined above in Equation 1

$V(x_i)$ = variance of random variable x_i as defined above in Equation 1

Note that the variance is simply the square of the standard deviation. Using the variance will simplify the mathematics that is described below. The contribution of each independent random variable to the overall variance in the function is:

$$V(Y \text{ from } x_i) = \left[\frac{\partial Y}{\partial x_i}\right]^2 V(x_i) \qquad \text{(Eq 4)}$$

The relative contribution of each term to the overall variance V(Y) is a measure of the importance in the uncertainty in the particular random variable, x_i. In effect, this is a sensitivity analysis combined with a uncertainty evaluation. The variance contribution combines the sensitivity in the answer to changes in the uncertain random variable, x_i, with a measure of the uncertainty in the random variable, x_i. The overall variance in Y is found by summing the sensitivity weighted variances from each random variable.

Numerical Estimate of the Sensitivity

Let us return to the fundamental definition of the derivative.

As before, set y = F(xi)

$$\frac{\partial y}{\partial x} = \lim_{\Delta xi \to 0} \left[\frac{F(xi + \Delta xi) - F(xi)}{\Delta xi}\right] \qquad \text{(Eq 5)}$$

We are interested in a numerical estimate of the derivative. Therefore, we will make the following approximation.

$$\frac{\partial y}{\partial x} \sim \left[\frac{F(xi + \Delta xi) - F(xi)}{\Delta xi}\right] \qquad \text{(Eq 6)}$$

Linear Functions

Let us examine the approximation for a simple linear function.

Let Y = AX + B (Eq 7)

What is the sensitivity of Y to a change in variable X? By simple calculus:

$$\frac{\partial y}{\partial x} = A \qquad \text{(Eq 8)}$$

We can also find the sensitivity using the numerical approximation as:

$$\frac{\partial y}{\partial x} \sim \left[\frac{F(xi+\Delta xi) - F(xi)}{\Delta xi}\right] \qquad \text{(Eq 9)}$$

$$\frac{\partial y}{\partial x} \sim \left[\frac{(A(xi+\Delta xi)+B)-(A(xi)+B)}{\Delta xi}\right] \qquad \text{(Eq 10)}$$

$$\frac{\partial y}{\partial x} \sim \left[\frac{(A(xi+\Delta xi))-(A(xi))}{\Delta xi}\right] \qquad \text{(Eq 11)}$$

$$\frac{\partial y}{\partial x} \sim \left[\frac{((A\,xi+A\,\Delta xi))-(A(xi))}{\Delta xi}\right] \qquad \text{(Eq 12)}$$

$$\frac{\partial y}{\partial x} \sim \left[\frac{A\,\Delta xi}{\Delta xi}\right] \qquad \text{(Eq 13)}$$

$$\frac{\partial y}{\partial x} = A \qquad \text{(Eq 14)}$$

Note that for a simple linear equation the numerical result for the sensitivity is exactly the same as the analytical expression. There is no restriction on the size of the perturbation (Δxi) used in the calculations. Previously, Freeman [Ref 7] has suggested using a 10% perturbation in xi to do the sensitivity calculations. However any size of perturbation will work and generate the correct sensitivity.

This is a general result. As long as the function can be expressed as a linear function of a number of variables, xi, the sensitivity can be directly calculated by a simple perturbation of the function. The resulting sensitivity will be exactly the same as would be obtained using analytical methods.

Non-Linear Expressions

In the quantification of a fault tree, the top event frequency is computed using failure rates of the component devices and time that measures the interval between system validations. In a minimal cut set of two or more basic events, the time variable may be the same. For example the test interval for a valve may be the same for a pressure transmitter. A minimal cut set involving both the valve and the pressure transmitter would create a 2rd order term in the test interval. If the test interval is uncertain, how do we compute the sensitivity of the top event frequency with respect to the test interval? We will now explore how to deal with this type of problem when completing the uncertainty analysis.

As before, let

$$y = F(xi) = Axi^{n} + B \qquad \text{(Eq 15)}$$

What is the sensitivity of Y to a change in variable X? By simple calculus:

$$\frac{\partial y}{\partial x} = nAxi^{n-1} \qquad \text{(Eq 16)}$$

Let us evaluate the numerical approximation for this case. If n=2, the function of interest is:

$y = F(xi) = Axi^2 + B$ (Eq 17)

From basic calculus, the exact analytical expression for the sensitivity is:

$$\frac{\partial y}{\partial x} = 2\,A(xi) \quad \text{(Eq 18)}$$

If xi = 1

$$\frac{\partial y}{\partial x} = 2\,A \quad \text{(Eq 19)}$$

Again the approximation of the derivative is:

$$\frac{\partial y}{\partial x} \sim \left[\frac{(A(xi+\Delta xi)^2+B)-(Axi+B)}{\Delta xi}\right] \quad \text{(Eq 20)}$$

$$\frac{\partial y}{\partial x} \sim \left[\frac{F(xi+\Delta xi)-F(xi)}{\Delta xi}\right] \quad \text{(Eq 21)}$$

Inserting the equation for F(xi) we obtain:

$$\frac{\partial y}{\partial x} \sim \left[\frac{(A(xi+\Delta xi)^2+B)-(Axi+B)}{\Delta xi}\right] \quad \text{(Eq 22)}$$

$$\frac{\partial y}{\partial x} \sim \left[\frac{(A\,(xi^2+2\Delta xi+\Delta xi^2)+B)-(Axi+B)}{\Delta xi}\right] \quad \text{(Eq 23)}$$

$$\frac{\partial y}{\partial x} \sim \left[\frac{A\,(xi^2+2\Delta xi+\Delta xi^2)-Axi}{\Delta xi}\right] \quad \text{(Eq 24)}$$

$$\frac{\partial y}{\partial x} \sim A\left[\frac{(xi^2+2\Delta xi+\Delta xi^2)-xi}{\Delta xi}\right] \quad \text{(Eq 25)}$$

Setting xi =1, and using a 10% perturbation as $\Delta xi = 0.1\,xi = 0.1$

$$\frac{\partial y}{\partial x} \sim A\left[\frac{(1^2+(2)(0.1)1+0.1^2)-1}{0.1\,(1)}\right] \quad \text{(Eq 26)}$$

$$\frac{\partial y}{\partial x} \sim A\left[\frac{(2)(0.1)+0.01}{0.1}\right] \quad \text{(Eq 27)}$$

$$\frac{\partial y}{\partial x} \sim A\left[\frac{0.21}{0.1}\right] \quad \text{(Eq 28)}$$

$$\frac{\partial y}{\partial x} = 2.1\,A \quad \text{(Eq 29)}$$

Note that the numerical estimate (Eq 29) is approximately 5% greater than the analytical value (Eq 19). As shown in Table 1, for higher order equations the error will increase as the size of the perturbation becomes more important. Table 2 shows the impact of reducing the perturbation size to 1%. For 10th order variables (say a minimal cut set with 10 basic events) the error is 5%. For almost all engineering evaluations, a maximum error of 5% should be adequate.

For cases where the test frequency, repair time, or mission time appears in a minimal cut set 3 or more times, I suggest that the following perturbation in time be used to compute the sensitivity of the top event:

$$\Delta time = 0.01\ time \qquad \text{(Eq 30)}$$

The sensitivity of the top event frequency to all uncertain variables can be computed using a perturbation of 1%.

$$\Delta xi = 0.01\ xi \qquad \text{(Eq 31)}$$

3. Fault Tree Analysis

Fault trees are often used to analyze potential failures of an engineered system. A fault tree is a failure logic diagram that shows the relationship of device or system failures leading to an accident. Fault tree analysis is covered in detail in the CCPS CPQRA book [Ref. 1]. A fault tree is constructed by starting with a top event or condition and asking the question:

What conditions lead to this outcome?

Each condition in the next layer down is then analyzed by asking the same question. What conditions lead to this outcome? The logic of how these intermediate conditions (termed intermediate events) are related is expressed using AND and OR logic gates. An AND gate output is true if and only if (iff) all of the inputs to the AND gate are all true simultaneously. An OR gate output is true iff one of the inputs to the OR gate is true.

Once the fault tree logic structure is created, numerical data can be used to calculate the frequency or probability of the top event of concern occurring. For fault trees where a basic event is found in multiple branches of the tree, such as power failure, the logic structure of the tree must be simplified to remove the impact of the repeated events on the top event of concern. This simplification is done using Boolean Algebra. Each resulting grouping of basic events can cause the top event to occur. These groupings of basic events are called minimal cut sets. As a simple example consider the Boolean Algebra representation of a fault tree :

$$T = BE11 + BE1 \bullet BE2 + BE1 \bullet BE3 + BE2 \bullet BE9 + BE3 \bullet BE7 \qquad \text{(Eq 32)}$$

Where:

T = Fault Tree Top Event

BE11 = Basic Event 11

BE1 = Basic Event 1

BE2 = Basic Event 2

BE3 = Basic Event 3

BE7 = Basic Event 7

BE9 = Basic Event 9

+ = Boolean Algebra Addition Symbol

• = Boolean Algebra Multiplication Symbol

For this example the top event T, the minimal cut sets are:

BE11

BE1 BE2

BE1 BE3

BE2 BE9

BE3 BE7

For example the existence of basic events BE1 and BE3 guarantees that the top event T will occur. Publically available standard computer software such as SAPHIRE [Ref. 8] may be used to find the minimal cut sets of a large fault tree.

The representation of the top event in equation 32 can now be used to compute the probability or frequency of occurrence of the top event. If the devices describe by the basic events are non-repairable and are tested after a period of time TI, the probability of top event, T, may be calculated as:

$$\text{Prob (T)} = \lambda_{11}\,TI/2 + (\lambda_1\,TI/2)(\lambda_2\,TI/2) + (\lambda_1\,TI/2)(\lambda_3\,TI/2) + (\lambda_2\,TI/2)(\lambda_9\,TI/2) + (\lambda_3\,TI/2)(\lambda_7\,TI/2) \quad \text{(Eq 33)}$$

Where:

λ_{11} = failure rate of device 11 in basic event BE11, failures per unit time

λ_1 = failure rate of device 1 in basic event BE1, failures per unit time

λ_2 = failure rate of device 2 in basic event BE5, failures per unit time

λ_3 = failure rate of device 3 in basic event BE3, failures per unit time

λ_7 = failure rate of device 7 in basic event BE7, failures per unit time

λ_9 = failure rate of device 9 in basic event BE9, failures per unit time

TI = test interval, time

Equation 33 assumes that the device failure rates are small and the product λ TI < 0.1. The other major assumption is that the values of all of the failure rates and test intervals are known. In reality the failure rate data are not known exactly with only a range or probability distribution representing the state of knowledge.

4. Example Problem

The following example is taken from the ISA technical report on fault tree analysis [Ref. 9]. This example should not be considered as a recommendation by either ISA or AIChE. The example is presented to allow for the demonstration of the methods that can be used in a fault tree study to quantify the uncertainty in the resulting calculated failure frequency or failure probability.

Consider the interlock on the intermediate storage tank T101 shown on Piping and Instrument Diagram in Figure 1. The interlock block diagram is shown in Figure 2. The interlock is intended to prevent an abnormal condition in the tank which could lead to a uncontrolled release of the tank contents. The process hazards analysis team has recommended that the interlock be designed to safety integrity level 2 (SIL 2) with a target probability of failure on demand (PFD) of no greater than 1E-2 or a risk reduction factor (RRF) of 100. Does the proposed interlock shown on Figure 1 meet the SIL 2 target? What is the uncertainty in the predicted PFD of the interlock?

A fault tree for the failure of this interlock is shown in Figure 3. A minimal cut set analysis has been completed and the resulting minimal cut sets are presented in Table 3. We can now write the Boolean Algebra equation that represents this fault tree as:

$$T = PE + TS1 \bullet TS2 + LS1 \bullet LS2 + FT1 \bullet FT2 + FT2 \bullet FT3 + FT1 \bullet FT3 + BV1 \bullet BV2 + BV1 \bullet SOL1 + BV2 \bullet SOL2 + SOL1 \bullet SOL2 + PT1 \bullet PT2 \quad \text{(Eq 34)}$$

Assuming that the interlock is non-repairable until tested at time TI and using lamda-time (λT) approximation of the failure rate, the Boolean representation of the fault tree may now be converted to a failure probability model as:

$$\begin{aligned} Prob(T) = {} & Prob(PE) + (\lambda_{TS1}\, TI/2)(\lambda_{TS2}\, TI/2) + (\lambda_{LS1}\, TI/2)(\lambda_{LS2}\, TI/2) + (\lambda_{FT1}\, TI/2)(\lambda_{FT2}\, TI/2) + \\ & (\lambda_{FT2}\, TI/2)(\lambda_{FT3} TI/2) + (\lambda_{FT1}\, TI/2)(\lambda_{FT3}\, TI/2) + (\lambda_{BV1}\, TI/2)(\lambda_{BV2}\, TI/2) + \\ & (\lambda_{BV1}\, TI/2)(\lambda_{SOL1} TI/2) + (\lambda_{BV2}\, TI/2)(\lambda_{SOL2}\, TI/2) + \\ & (\lambda_{SOL1}\, TI/2)(\lambda_{SOL2}\, TI/2) + (\lambda_{PT1}\, TI/2)(\lambda_{PT2}\, TI/2) \end{aligned} \quad \text{(Eq 35)}$$

Where:

Prob(PE) = failure probability of electronic logic solver, assumed to be constant at 5E-3.

λ_{TS1} = failure rate of temperature switch 1, failures per unit time

λ_{TS2} = failure rate of temperature switch 2, failures per unit time

λ_{LS1} = failure rate of level switch 1, failures per unit time

λ_{LS2} = failure rate of level switch 2, failures per unit time

λ_{FT1} = failure rate of flow transmitter 1, failures per unit time

λ_{FT2} = failure rate of flow transmitter 2, failures per unit time

λ_{FT3} = failure rate of flow transmitter 3, failures per unit time

λ_{BV1} = failure rate of block valve 1, failures per unit time

λ_{BV2} = failure rate of block valve 2, failures per unit time

λ_{SOL1} = failure rate of solenoid valve 1, failures per unit time

λ_{SOL2} = failure rate of solenoid valve 2, failures per unit time

λ_{PT1} = failure rate of pressure transmitter 1, failures per unit time

λ_{PT2} = failure rate of pressure transmitter 2, failures per unit time

TI = test interval, time

The electronic logic solver PE is assumed to have a constant failure probability of 5E-3.

Failure rate data for the other devices are present in Table 4. The mode (most likely value) of the failure rate is set equal to the point value taken from the ISA example [Ref. 9]. The triangular probability distribution (Appendix A) was assumed to represent the failure rate data for the equipment items. The upper and lower limits are based on the failure rate data of Appendix 4 of Smith [Ref. 10].

Using the device failure rates represented by the Table 4 column labeled as the mode and the minimal cut sets presented in Table 3, the ISA technical report [Ref.9, page 35] computes an interlock PFD as 7.5E-3 (RRF of 133). This would satisfy the requirement for a SIL 2 interlock {PFD of 1E-2 or a RRF of 100}. Now what is the uncertainty in the interlock PFD?

The first thing to remember is that the PFD should be calculated based on the mean value of the failure rate lambda (λ), not the mode of the failure rate. For a data set, the mean is the best single point representation of the data set. Table 5 presents the PFD of each minimal cut set using the mean of the device failure rate lambda (λ). The total of the PFD becomes 1.47E-2 (RRF of 68). This would not satisfy a SIL 2 interlock requirement.

Now we will compute the variance of the predicted probability of top event, T. To compute the variance of the probability of the top event, T, we will need to compute the sensitivity of the top event to each of the variables.

Sensitivity Using Numerical Perturbation Calculation

We will use the ISA interlock example previously analyzed by Freeman (Ref. 7). Using public data sources Freeman found that the probability of failure on demand (PFD) was 1.49E-2. To use the perturbation method for the evaluation of the sensitivities requires evaluating the impact of small changes in the uncertain input parameters. To enable these calculations to be completed numbers using 8 significant digits were used. The calculations were completed using Microsoft Excel which uses double-precision floating-point arithmetic compliant with IEEE 754 specification. Excel nominally carries 15 significant figures in calculations. We will complete the calculations with Excel displaying 5 significant digits and round the result at the end of the calculations to 4 significant digits. We will illustrate the calculations by looking at the Flow Transmitter, FT1.

a. First we recalculate the PFD of the example and find it to be 1.4594E-02. This is the base number that we will use in the perturbation calculations.
b. We now make a small change in mean failure rate of FT1. The mean failure rate is 8E-6 failures/hour. A 1% perturbation is used or an increase of 0.08E-6 failures/hour.

c. The perturbed value of the failure rate of FT1 becomes (8 + 0.08) * 1E-6 failures/hour or 8.08E-6 failures/hour.
d. The perturbed value of the failure rate of FT1 is now used to re-calculate the PFD of the interlock. The resulting re-calculation finds that the PFD is 1.4618E-02
e. The change in the PFD from the base number due to the perturbation of the failure rate of FT1 is now computed as:

   delta PFD = 1.4618E-02 - 1.4594E-02 = 2.4556E-05

f. The sensitivity of the PFD to a change in the failure rate of FT1 is now computed by dividing the change in the PFD by the change FT1 failure rate as:

   sensitivity = (2.4556E-05)/(0.08E-6 failures/hour) = 307.0 hours
   Result is rounded to 4 significant digits.

We now note that in the paper by Freeman (Ref. 7) that equation 32 presents the sensitivity for FT1 as:

$$\frac{\partial Prob(T)}{\partial \lambda FT1} = 307.0 \text{ hr} \qquad \text{(Eq 32 of Ref 7)}$$

The answers are exactly the same. The sensitivity calculations for all of the devices are presented in Table 6. Note that in every case the sensitivity found using the analytical expression and the numerical estimate are the same. This is a general result which will be true for all basic event equipment failure rates found in a fault tree minimal cut set.

Once found, the sensitivities can now be used to compute the variance contribution of each particular device to the overall variance of the top event PFD in the same manner as Freeman previously presented.

The sensitivities of the top event probability are summarized in Table 6. We may now compute the variance contribution of each uncertain variable using Equations 1 – 4 in the methodology section of this paper. The relative variance contribution is expressed as a percent of the total variance of the top event probability. The top event variance is found to be 8.24E-6 (failures per hour)^2 or equivalently the top event standard deviation is 2.78E-3 failures per hour.

In this example the variance (measure of uncertainty) is dominated by the flow transmitters. The flow transmitters account for almost 2/3 of the uncertainty in the top event probability. Failures of the flow transmitters appear in 3 of the minimal cut sets. Since the uncertainty in the transmitter failure rates is large, the flow transmitters should contribute a significant amount to the variance of the fault tree top event.

We can use the variance in the top event to define the likelihood of achieving SIL 2 performance of the interlock. The calculations are done using the normal probability distribution. The normal distribution is tabulated as the standard normal distribution using a normalization factor Z.

The standard normal factor, Z, [Ref. 11] is defined as:

$$Z = \left[\frac{xi - E(x)}{\sigma}\right] \qquad \text{(Eq 36)}$$

Where:

σ = standard deviation. Note that the variance of a random variable is the square of the standard deviation of the random variable.

$E(x)$ = Expected value of the random variable x_i

For the example, the expected value of the probability of the top event of concern was calculated as

$$E[Prob(T)] = 1.47E\text{-}2 \quad \text{(Eq 37)}$$

From Table 4, the standard deviation in the probability of the top event is

$$\sigma_T = 2.78E\text{-}3 \quad \text{(Eq 38)}$$

The target PFD for a SIL 2 interlock must not be worse than 1E-2 (RRF of 100).

We compute the normal distribution Z factor at the SIL 2 target value as

$$Z = (0.01 - 0.0147)/2.78E\text{-}3 \quad \text{(Eq 39)}$$

$$Z = -1.69065 \quad \text{(Eq 40)}$$

Tabulations of standard normal distribution are presented in most statistics books such as Meyer [Ref. 11]. For this value of Z, the corresponding probability that the interlock will achieve SIL 2 performance is 4.55 percent. We can also compute the risk reduction that the proposed interlock design will achieve.

What is the interlock probability of failure on demand (PFD) that we can be 95% certain that the interlock will provide? At the 95% level, the corresponding Z factor is 1.65 [Ref 11]. We now compute the corresponding PFD of the interlock that corresponds to this Z factor.

$$Z = \left[\frac{x95\% - E[Prob(T)]}{\sigma}\right] \quad \text{(Eq 41)}$$

In this case:

$x_{95\%}$ = the PFD that we can be confident that the interlock will achieve

$$\sigma_T = 2.78E\text{-}3 \quad \text{(Eq 42)}$$

$$E[Prob(T)] = 1.47E\text{-}2 \quad \text{(Eq 43)}$$

$$Z = 1.65 \quad \text{(Eq 44)}$$

Rearranging equation 40 to find $x_{95\%}$we obtain:

$$x_{95\%} = E[Prob(T)] + Z\,\sigma_T \quad \text{(Eq 45)}$$

$$x_{95\%} = 1.47E\text{-}2 + (1.65)\,(2.78E\text{-}3) = 0.0193 \quad \text{(Eq 46)}$$

or a 95% certain risk reduction factor of

$$RRF = 1/PFD = 1/0.0193 = 52 \quad \text{(Eq 47)}$$

By the same logic there is 5% chance that the risk reduction is a PFD of 0.01 or an RRF of 99. In summary, this interlock is likely to perform as a mid-range SIL 1 interlock, not the SIL 2 interlock desired by process hazards analysis team.

5. Methodology for Fault Tree Uncertainty Analysis

We can now generalize the methods used in the above example to a proposed method for the evaluation of the uncertainty in a fault tree. The following step-by-step procedure is modified from that that previously published by Freeman (Ref 7) to indicate the use of numerical methods to compute the sensitivity. Step 8 shown in *italics* indicates the modification..

1. Create the fault tree using standard methods outline in the CPQRA book [Ref 1].
2. Determine minimal cut sets by hand or using standard computer software such as SAPHIRE [Ref 12].
3. Define the needed failure rate data for each basic event.
4. Define those basic events that are considered to be uncertain.
5. For the basic events that are uncertain, define the probability distribution and associated parameters needed to numerically define the probability distribution. Appendix B of this paper presents the needed description of 4 commonly used probability distributions. In many cases the triangular distribution will be selected. The minimum, maximum and mode (most likely) parameters will be needed for the triangular distribution.
6. Compute the mean of each of the uncertain variable
7. Compute the probability or frequency of the fault tree top event of interest using the mean value for each uncertain variable.
8. *Compute the sensitivity of the top event probability or frequency using the numerical methods outlined in this paper.*
9. Compute the variance of each of the uncertain variables
10. Compute the variance contribution of for each of the uncertain variables to the top event probability or frequency using equation 6.
11. Compute the total variance of the top event of interest probability or frequency by summing all of the contributions determined in Step 10.
12. Compute the variance contribution percent of each uncertain variable by dividing the variable contribution (Step 10) by the total variance of the top event of interest (Step 11).
13. Define the level of risk the project is willing to take. What chance will the project management accept for potential failure of the interlock to achieve the desired risk reduction? In the above example, I have used a 5% risk of failure or a 95% certainty that the interlock will achieve a risk reduction factor of 52. In the example, there is a very low probability that the interlock will achieve a SIL 2 PFD of 0.01 or an RRF of 100.

Alternately you can report the 90% range for the top event of interest. For the example the 90% range starts at the 5% RRF of 99 and there is a 95% certainty that that an RRF of at least 52 will be achieved. The expected risk reduction for this interlock is an RRF of 68.

6. Conclusions

A general procedure has been presented to quantify the uncertainty in calculations of failure frequencies using the fault tree methodology. The uncertainty analysis procedure is based on the application of propagation of error and variance contribution analysis techniques to the minimal cut sets created during a fault tree study. The procedure is simple and can be incorporated into standard fault tree analysis programs such as SAPHIRE [Ref 8]. This procedure is based on perturbation calculation of the frequency of the top event from changes in the uncertain variables. Incorporation of a numerical perturbation method into standard fault tree software such as SAPHIRE would be easy to implement and would provide the means for evaluation of the uncertainty in the quantified fault tree results.

For the example interlock presented in this paper, the uncertainty in the device failure rates degrades the predicted interlock SIL from meeting SIL 2 requirements to becoming a mid range SIL 1 interlock. This is a general result. Uncertainty in design parameters will reduce the likelihood that the equipment will achieve desired results. The design engineer has two choices:

- Accept the conclusion that the interlock will not perform as SIL 2
- Re-design the interlock using more reliable components to achieve the SIL 2 target

The variance contribution analysis shown in Table 7 indicates that the major contributor to the uncertainty in the interlock performance is the Flow Transmitters. Flow transmitters certified to SIL-2 or SIL-3 performance with lower failure rates could improve the calculated PFD of the interlock.

7. References

1. Center for Chemical Process Safety , *Guidelines for Chemical Process Quantitative Risk Analysis – 2rd edition*, American Institute of Chemical Engineers, New York, 2000.
2. Frank P Lees, Loss Prevention in the Process Industries – 2rd Edition, Butterworth, ISBN 0-7506-1547-8, Oxford, UK, 1996
3. US Department of Labor, Occupational Safety and Health Administration, CFR 1910.119 - Process Safety Management of Highly Hazardous Chemicals, 1910.119(e)(2)(vi).
4. Edward B. Haugen, *Probabilistic Approaches to Design*, John Wiley, New York, 1968
5. Raymond. Freeman and Angela Summers, "Evaluation of Uncertainty in Safety Integrity Level Calculations," *Process Safety Progress,* Published online in Wiley Online Library (wileyonlinelibrary.com) DOI 10.1002/prs. 11805, 2016
6. Raymond. Freeman, "General Method for Uncertainty Evaluation of Safety Integrity Level Calculations – Part 2 Analytical Methods," *Process Safety Process*, Published online in Wiley Online Library (wileyonlinelibrary.com) DOI 10.1002/prs 111915, October 2017
7. Raymond. Freeman ,"Error Propagation and Uncertainty Analysis: Application to Fault Tree Analysis", Process Safety Progress, Published Wiley Online Library (wileyonlinelibrary.com/journal/prs) DOI: https://doi.org/10.1002/prs.12080, June 29, 2019

8. US Nuclear Regulatory Commission, "Systems Analysis Programs for Hands-on Integrated Reliability Evaluations (SAPHIRE) – Version 8," NUREG/CR-7039, Vol 3, June 2011.
9. ISA, "Safety Instrumented Functions (SIF) – Safety Integrity Level (SIL) Evaluation Techniques Part 3: Determining the SIL of a SIF via Fault Tree Analysis, ISA Technical Report ISA-TR-84.00.02-2002 – Part 3, Approved 17 June 2002.
10. David J. Smith, *Reliability Maintainability and Risk – 8th Edition*, ISBN 978-0-08-096902-2, Elsevier, London, 2011
11. Paul L. Meyer, Introductory Probability and Statistical Applications, John Wiley, New York, 1972

Table 1. Comparison of Analytical Sensitivity with Numerical Approximation with Perturbation of 10% in xi for the equation: $y = F(xi) = Axi^n + B$

Order of Equation n	Equation for Analytical Sensitivity $\frac{\partial y}{\partial x} = nAxi^{n-1}$	Sensitivity with xi = 1 and A=1	Numerical Sensitivity with $\Delta xi = 0.1\, xi$	Error %
1	$\frac{\partial y}{\partial x} = 1A(xi)^0$	1	1	0
2	$\frac{\partial y}{\partial x} = 2A(xi)^1$	2	2.1	5
3	$\frac{\partial y}{\partial x} = 3A(xi)^2$	3	3.31	-10%
4	$\frac{\partial y}{\partial x} = 4A(xi)^3$	4	4.64	-16%
5	$\frac{\partial y}{\partial x} = 5A(xi)^4$	5	6.11	-22%
6	$\frac{\partial y}{\partial x} = 6A(xi)^5$	6	7.72	-29%
7	$\frac{\partial y}{\partial x} = 7A(xi)^6$	7	9.49	-36%
8	$\frac{\partial y}{\partial x} = 8A(xi)^7$	8	11.44	-43%
9	$\frac{\partial y}{\partial x} = 9A(xi)^8$	9	13.58	-51%
10	$\frac{\partial y}{\partial x} = 10A(xi)^9$	10	15.94	-59%

Error = (Analytical- Numerical)/Analytical

Table 2. Comparison of Analytical Sensitivity with Numerical Approximation with Perturbation of 1% in xi for the equation: $y = F(xi) = Axi^n + B$

Order of Equation n	Equation for Analytical Sensitivity $\frac{\partial y}{\partial x} = nAxi^{n-1}$	Sensitivity with xi = 1 and A=1	Numerical Sensitivity with $\Delta xi = 0.1\ xi$	Error %
1	$\frac{\partial y}{\partial x} = 1A(xi)^0$	1	1	0
2	$\frac{\partial y}{\partial x} = 2A(xi)^1$	2	2.01	-1%
3	$\frac{\partial y}{\partial x} = 3A(xi)^2$	3	3.03	-1%
4	$\frac{\partial y}{\partial x} = 4A(xi)^3$	4	4.06	-2%
5	$\frac{\partial y}{\partial x} = 5A(xi)^4$	5	5.10	-2%
6	$\frac{\partial y}{\partial x} = 6A(xi)^5$	6	6.15	-3%
7	$\frac{\partial y}{\partial x} = 7A(xi)^6$	7	7.21	-3%
8	$\frac{\partial y}{\partial x} = 8A(xi)^7$	8	8.29	-4%
9	$\frac{\partial y}{\partial x} = 9A(xi)^8$	9	9.37	-4%
10	$\frac{\partial y}{\partial x} = 10A(xi)^9$	10	10.46	-5%

Error = (Analytical- Numerical)/Analytical

Table 3. Basic Events in Minimal Cut Sets For Example

Cut Set No	BE 1	BE 2
1	PE	-
2	TS1	TS2
3	LS1	LS2
4*	FT1	FT2
5*	FT2	FT3
6*	FT1	FT3
7*	BV1	BV2
8*	BV1	SOL1
9*	BV2	SOL2
10*	SOL1	SOL2
11	PT1	PT2

* Minimal cut Set with same basic event as another cut set

Table 4. Lamda (λ) Failure Rate Data With Uncertainty Limits

Device	Minimum Failures per 1E6 Hours	Mode Failures per 1E6 Hours	Maximum Failures per 1E6 Hours	Mean Failures per 1E6 Hours	Lamda Variance (Failures per 1E6 Hours)^2
Flow Transmitters FT1, FT2 and FT3	1.0	2.9	20.0	8.0	18.3
Pressure Transmitters PT1 and PT2	1.0	2.3	20.0	7.8	18.8
Temperature Switch TS1 and TS2	3.0	7.6	20.0	10.2	12.9
Level Switch LS1 and LS2	2.0	4.6	20.0	8.9	15.8
Block Valves BV1 and BV2	0.2	2.3	10.0	4.2	4.4
Solenoid Valves SOL1 and SOL2	1.0	2.3	8.0	3.8	2.3

Table 5. Probability of Minimal Cut Sets Based on Mean Failure Rates

Cut Set No	BE 1	BE 2	PFD based on mean λ
1	PE	-	5.00E-03
2	TS1	TS2	2.00E-03
3	LS1	LS2	1.52E-03
4	FT1	FT2	1.23E-03
5	FT2	FT3	1.23E-03
6	FT1	FT3	1.23E-03
7	BV1	BV2	3.38E-04
8	BV1	SOL1	3.38E-04
9	BV2	SOL2	3.38E-04
10	SOL1	SOL2	2.77E-04
11	PT1	PT2	1.17E-03

Table 6. Sensitivity of Example System PFD to Basic Event Failure Rates Based on a 1% Perturbation

Basic Event Label	Device Type	Mean Failures per 1E6 Hours	Failure Perturbation of 1% of Mean Failures per 1E6 Hours	Perturbed Failure Rate Failures per 1E6 Hours	Total System PFD Using Perturbed Failure Rate of Basic Event	Change in Total System PFD	Sensitivity of PFD to Change in Failure Rate Hours
FT1	Flow Transmitter	8	0.08	8.08	1.461848E-02	2.455603E-05	307.0
FT2	Flow Transmitter	8	0.08	8.08	1.461848E-02	2.455603E-05	307.0
FT3	Flow Transmitter	8	0.08	8.08	1.461848E-02	2.455603E-05	307.0
PT1	Pressure Transmitter	7.8	0.078	7.878	1.460560E-02	1.167179E-05	149.6
PT2	Pressure Transmitter	7.8	0.078	7.878	1.460560E-02	1.167179E-05	149.6
TS1	Temperature Switch	10.2	0.102	10.302	1.461389E-02	1.995945E-05	195.7
TS2	Temperature Switch	10.2	0.102	10.302	1.461389E-02	1.995945E-05	195.7
LS1	Level Switch	8.9	0.089	8.989	1.460912E-02	1.519596E-05	170.7
LS2	Level Switch	8.9	0.089	8.989	1.460912E-02	1.519596E-05	170.7
BV1	Block Valves	4.2	0.042	4.242	1.460037E-02	6.445958E-06	153.5
BV2	Block Valves	4.2	0.042	4.242	1.460037E-02	6.445958E-06	153.5
SOL1	Solenoid Valves	3.8	0.038	3.838	1.459976E-02	5.832058E-06	153.5
SOL2	Solenoid Valves	3.8	0.038	3.838	1.459976E-02	5.832058E-06	153.5

Table 7. Variance of Top Event Probability Due to Uncertain Device Failure Rates

Device No	Device Type	Label	Sensitivity of Top Event Probability to Device Failure Rate $Hour^{-1}$	Variance of Device Failure Rate $(Failures\ per\ 1E6\ Hour)^2$	Variance Contribution to Top Event Variance	Contribution to Top Event Variance %
1	Logic Solver	PE	-	0	0.00E+00	0%
2	Temperature Switch	TS1	195.7	12.9	4.94E-07	6%
3	Temperature Switch	TS2	195.7	12.9	4.94E-07	6%
4	Level Switch	LS1	170.7	15.8	4.60E-07	6%
5	Level Switch	LS2	170.7	15.8	4.60E-07	6%
6	Flow Transmitter	FT1	307	18.3	1.72E-06	21%
7	Flow Transmitter	FT2	307	18.3	1.72E-06	21%
8	Flow Transmitter	FT3	307	18.3	1.72E-06	21%
9	Block Valve	BV1	153.5	4.4	1.04E-07	1%
10	Block Valve	BV2	153.5	4.4	1.04E-07	1%
11	Pressure Transmitter	PT1	149.6	18.8	4.21E-07	5%
12	Pressure Transmitter	PT2	149.6	18.8	4.21E-07	5%
13	Solenoid valve	SOL1	153.5	2.3	5.42E-08	1%
14	Solenoid valve	SOL2	153.5	2.3	5.42E-08	1%

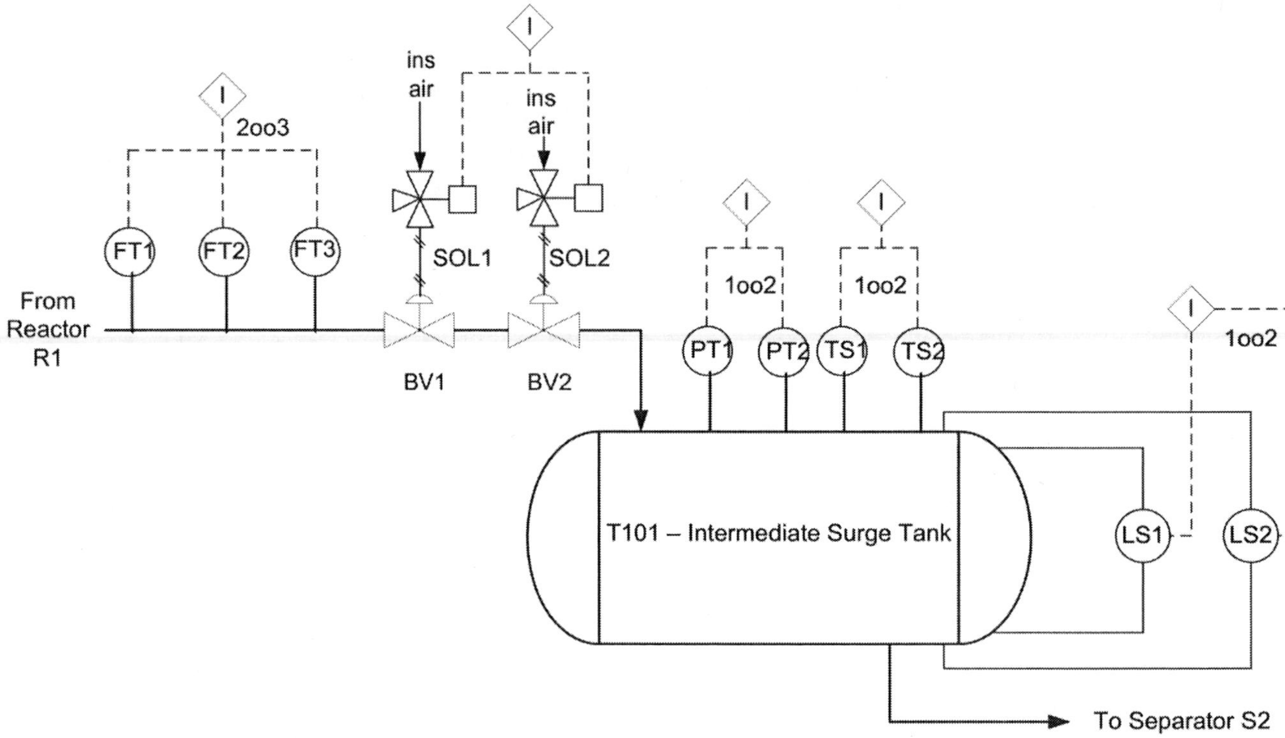

Figure 1. P&ID Diagram for Example

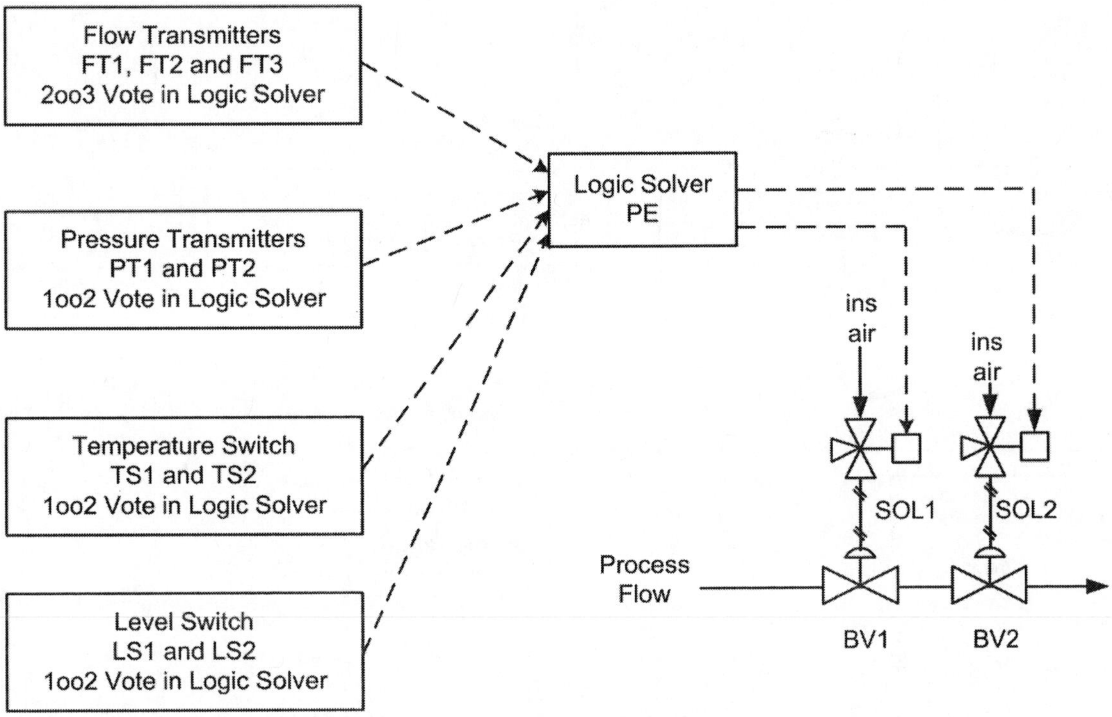

Figure 2. Interlock Block Diagram

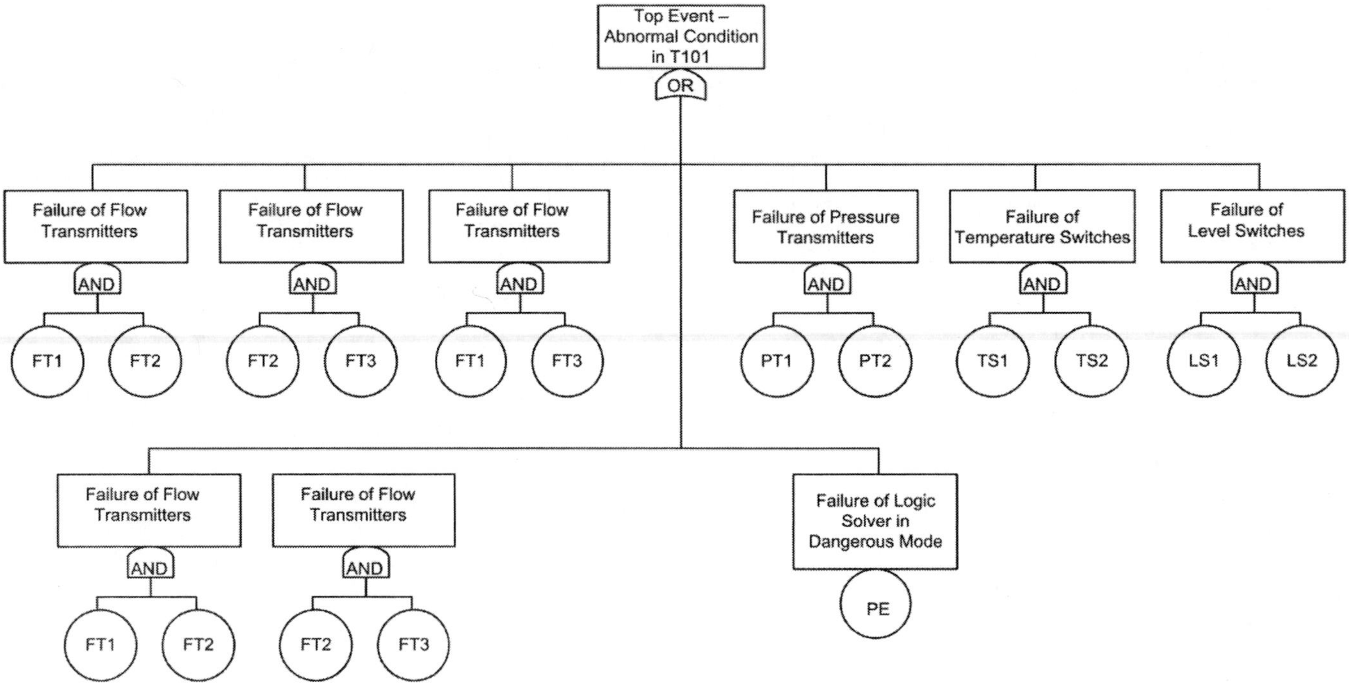

Figure 3. Fault Tree for Example Interlock

Appendix A – Commonly Encounter Probability Distributions

There are four probability distributions used to represent the failure rate of equipment items. These are:

1. Uniform Distribution
2. Triangular Distribution
3. Normal Distribution
4. Log-Normal Distribution

This appendix reviews the properties of these four probability distributions.

1. Uniform Distribution

The uniform distribution [Ref. A-1, pg 74; A-2, pg 687-688] is used when a variable is only known within a defined range. Figure A-1 presents a plot of the probability density function (pdf) for the uniform distribution. The distribution is defined using two parameters, A and B.

Where:

A = minimum value of the random variable

B = maximum value of the random variable

The mean or expected value of a random variable, x, that follows the uniform distribution is:

Mean = Expected Value – $E(x) = (A+B)/2$ (Eq A-1)

The variance of a random variable, x, that follows the uniform distribution is:

Variance = $V(x) = (B-A)^2/12$ (Eq A-2)

The standard deviation of a random variable, x, that follows the uniform distribution is:

Standard Deviation = $(\text{Variance})^{1/2} = (B-A)/(12)^{1/2}$ (Eq A-3)

2. Triangular Distribution

The triangular distribution [Ref. A-2, pg 686-687] is used when a variable is known to lie within a defined range and a "best guess" or estimate can be made as to the most likely value to the variable. Figure A-2 presents a plot of the probability density function (pdf) for the triangular distribution. The distribution is defined using three parameters, A. B and C.

Where:

A = minimum value of the random variable

B = maximum value of the random variable

C = most likely (mode) of the random variable

The mean or expected value of a random variable, x, that follows the triangular distribution is:

Mean = Expected Value – $E(x) = (A+B+C)/3$ (Eq A-4)

The variance of a random variable, x, that follows the triangular distribution is:

Variance = V(x) = $[A^2 + B^2 + C^2 - AB - AC - BC]/18$ (Eq A-5)

The standard deviation of a random variable, x, that follows the triangular distribution is:

Standard Deviation = $(\text{Variance})^{1/2} = \{[A^2 + B^2 + C^2 - AB - AC - BC]/18\}^{1/2}$ (Eq A-6)

3. Normal distribution

The normal distribution [Ref A-2, pg 665-666] is used when a variable when data analysis finds the normal distribution is the best model to describe the spread in the measured variable. Figure A-3 presents a plot of the probability density function (pdf) for the normal distribution. The distribution is defined using two parameters, μ and σ.

Where:

μ = mean of the measured data for variable

σ = standard deviation of the measured data for variable

The expected value of a random variable, x, that follows the normal distribution is:

Expected value = mean = $E(x) = \mu$ (Eq A-7)

The variance of a random variable, x, that follows the normal distribution is:

Variance = square of standard deviation = $V(x) = \sigma^2$ (Eq A-8)

The standard deviation of a random variable, x, that follows the normal distribution is:

Standard deviation = σ (Eq A-9)

4. Lognormal Distribution

The lognormal distribution [Ref. A- 2, pg 658-659] is used when a variable when data analysis finds the lognormal distribution is the best model to describe the spread in the measured variable. Figure A-4 presents a plot of the probability density function (pdf) for the lognormal distribution. Many variables in nature are lognormally distributed (for example the height of people). The log normal distribution is used when the variable of interest has a known physical lower limit of zero. A variable, x, is lognormally distributed if ln(x) is normally distributed. The lognormal distribution is defined using two parameters, μ_y and σ_y.

Let the original data be defined by the variable x with mean and standard deviation as:

μ_x = mean of the measured data for variable x

σ_x = standard deviation of the measured data for variable x

Define a new variable y as:

$y = \text{Ln}(x)$ (Eq A-10)

Then variable x is said to be lognormally distributed if y is normally distributed. We can now write the mean and standard deviation of y (μ_y and σ_y.) [Ref. A-3, A-4 and A-5] as

$$\mu_y = Ln\left[\mu_x^2 /((\sigma_x^2 + \mu_x^2)^{1/2})\right] \qquad \text{(Eq A-11)}$$

$$\sigma_y = \left[Ln(\sigma_x^2 / \mu_x^2 + 1)\right]^{1/2} \qquad \text{(Eq A-12)}$$

Where:

Ln is the natural logarithm (base e) of the argument

Appendix A References:

A-1. P. L. Meyer, *Introductory Probability and Statistical Applications*, 2nd Edition, Addison-Wesley, Reading Mass, Library of Congress Catalog no. 75-104971, 1972, pp. 74

A-2. D. Vose, *Risk Analysis–A Quantitative Guide*, 3rd edition, Wiley, Chichester, West Sussex, England, ISBN 978-0-470-51284-5, 2008, pp. 686-687

A-3. "Lognormal mean and variance - MATLAB lognstat". www.mathworks.com. Retrieved 27 December 2018

A-4. M. Mood, F. A. Graybill, and D. C. Boes. *Introduction to the Theory of Statistics.* 3rd ed., New York: McGraw-Hill, ISBN 0-07-042864-6, 1974. pp. 117.

A-5. E.L. Crow and K. Shimizu, Lognormal Distributions – Theory and Applications, Marcel Dekker, New York, ISBN 0-8247-7803-0, 1988, pp. 9

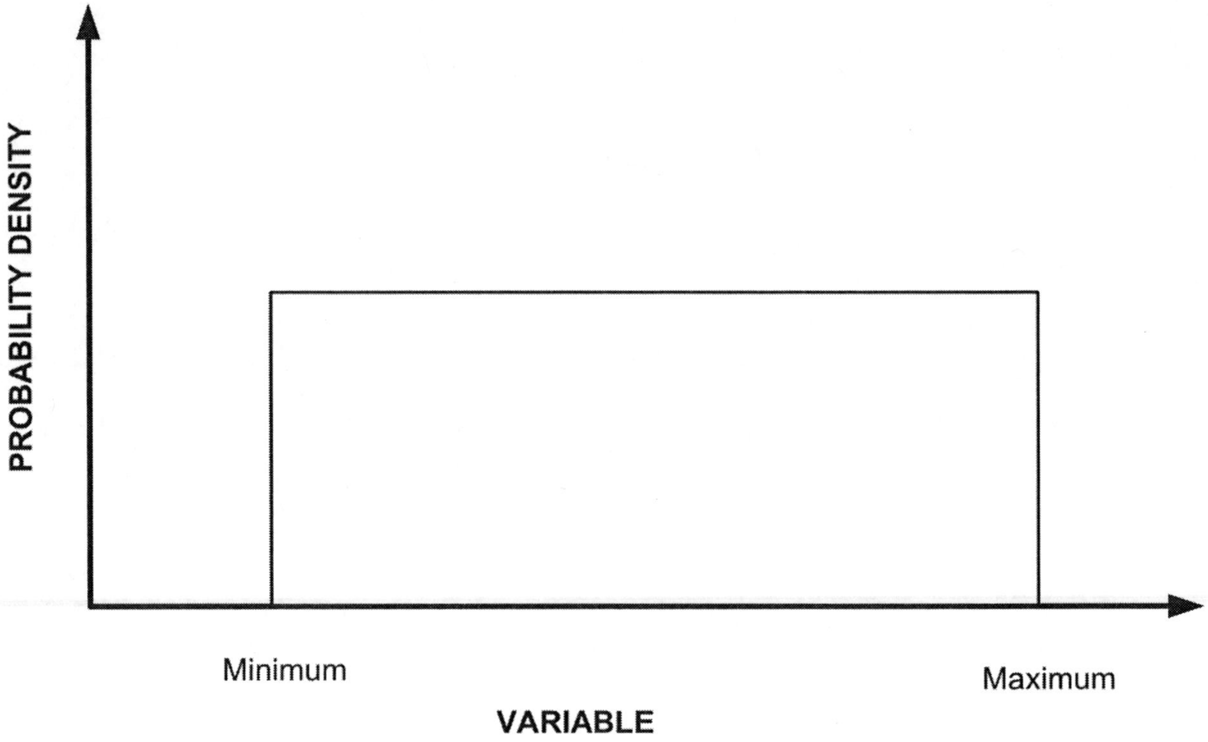

Figure A-1. Uniform Probability Distribution

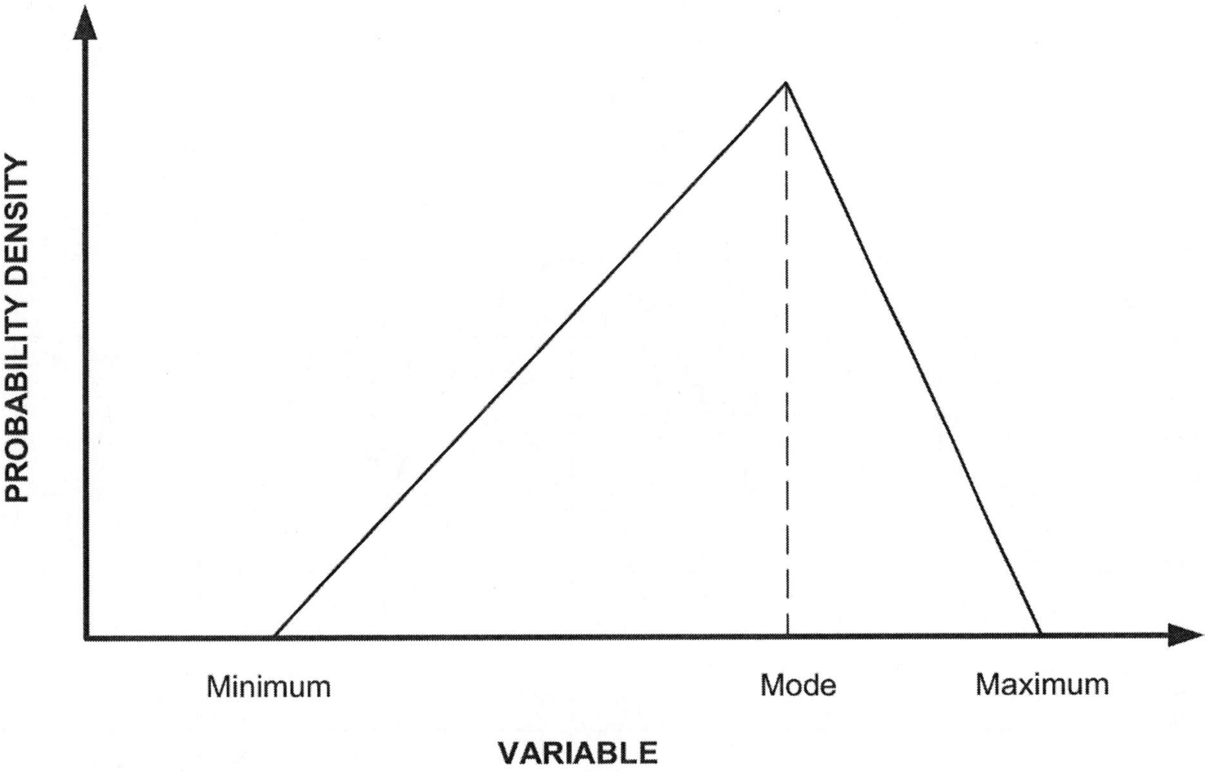

Figure A-2. Triangular Probability Distribution

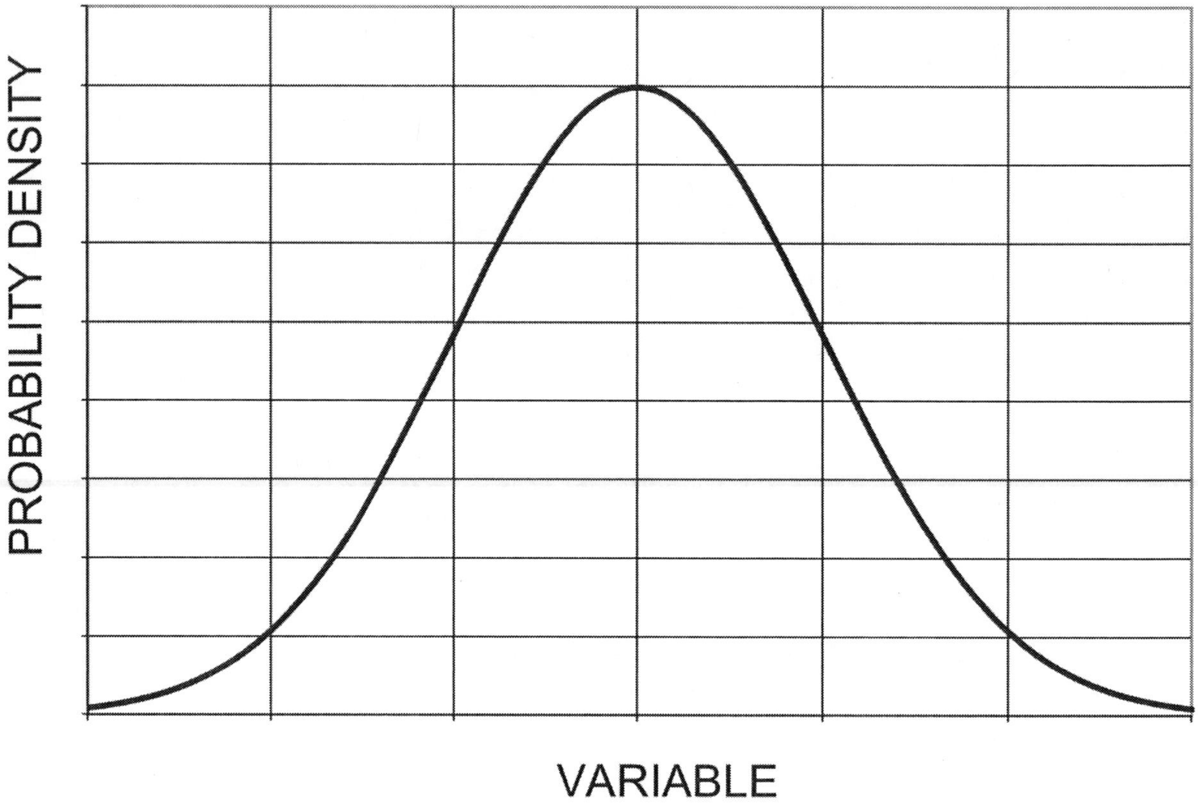

Figure A-3. Normal Probability Distribution

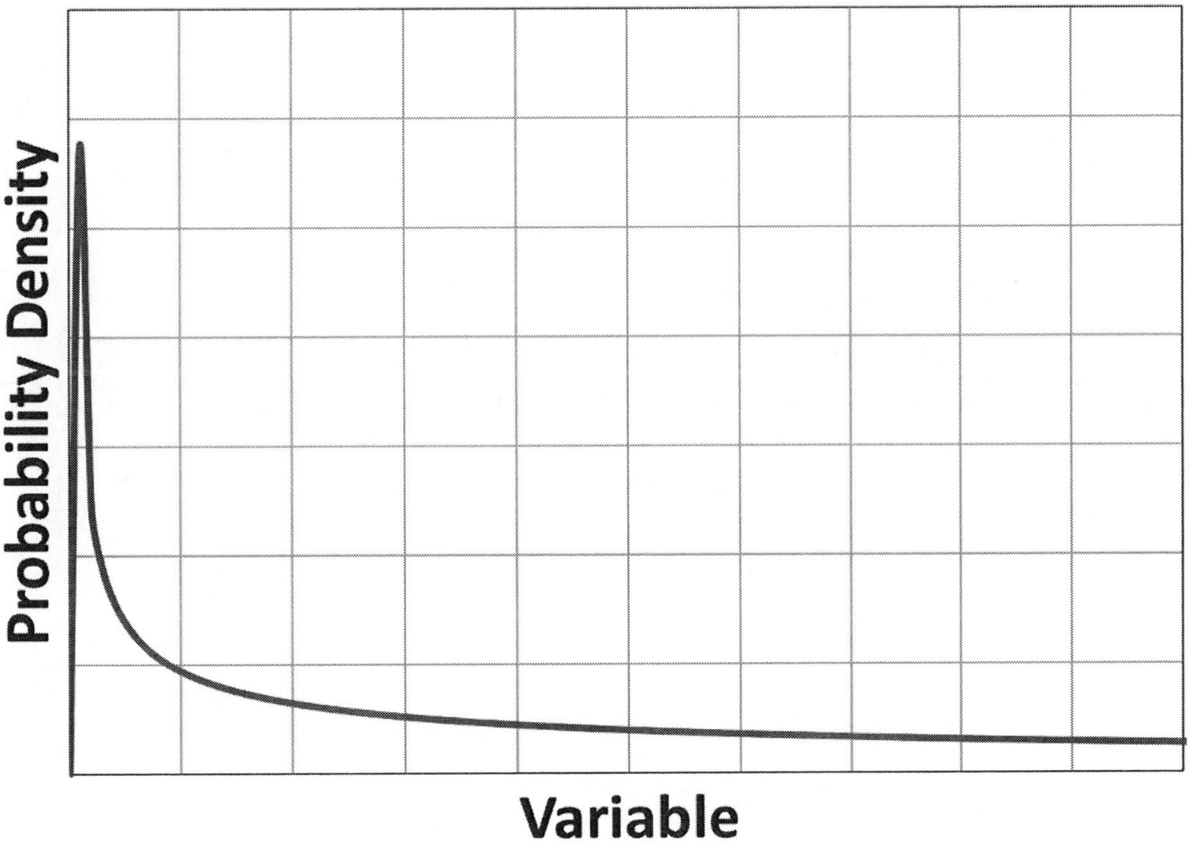

Figure A-4. Lognormal Probability Distribution

590424 Making the Most of Your PHA

Abstract:

A Process Hazard Analysis (PHA) is required to address OSHA Process Safety Management (PSM) and EPA Risk Management Program (RMP) regulations. Understanding the potential process hazard is widely recognized as a foundational element of PSM and RMP, and in addition to simply meeting regulatory requirements, can be a critical mechanism for flushing out potential vulnerabilities early in the design process for capital projects. Although an important activity, PHA is also recognized as a sizeable investment in time and energy for an Owner/Operator. So why not make the most of it? Other important activities at plants such as Mechanical Integrity, Alarm Rationalization, and Management of Change are able to stay focused and possibly be streamlined (i.e., reduced cost) based on the ability to use the results of a quality PHA having the necessary information and usability of documentation. So why not structure the PHA and documentation to be able to capitalize on potentially significant cost-savings and possible program integration?
The thrust of this paper and conference presentation will be to identify tips for conducting and documenting the PHA to maximize its ability to be used for supporting other important programs such as MOC, MI, and Alarm Rationalization. The objective will be to allow PHA Coordinators to be able to readily apply these tips to bring about improvements and cost-savings on these interrelated programs.

589814 Theoretical Analysis of the Viability of Isopropanol-Cyclohexane Solvent Systems for Biomatter Lipid Extraction

Abstract:

A theoretical alternative solvent combination for the extraction of lipids from biomatter is the Water, Isopropanol, and Cyclohexane instead of the more common combination used in the Bligh and Dyer Method (Water, Methanol, and Chloroform). While the Bligh and Dyer Method is considered effective, toxicological concerns have been raised over its use of both Chloroform and Methanol. The alternative method utilizes solvents that do not pose as high as a toxicological concern while theoretically maintaining a similar level of practical viability. The alternative method's extraction efficacy was judged against the performance of the Bligh and Dyer method using the simulation software Aspen Plus V10®. The equilibrium data generated from Aspen Plus V10® was then graphed into tertiary diagrams for data analysis using the graphing software SigmaPlot.

592166 Techno-Economic Analysis of the Conversion of CH4 and CO2 Intoacetic Acid Using Microwave Plasma

Abstract:

Methane and carbon dioxide together represent 90% of greenhouse gases [1], and one way of diminishing their emissions is through the development of pathways that convert them into added value chemicals. Acetic acid is one of such chemicals, and efforts at the laboratory scale have been carried out to obtain alternative ways of producing it. However, there are very few studies that evaluate the possibility of implementing these processes at a larger scale. In this paper, techno-economic and sustainability analyses were used to assess the feasibility of upscaling two pathways to produce acetic acid: (1) by direct conversion of methane and carbon dioxide and (2) by indirect conversion, using syngas (obtained from dry methane reforming) as an intermediate. For the direct method a microwave non-catalytic plasma reactor was considered as the main driver in the process simulation [2]. For the indirect method, a zeolite catalytical single step reaction unit was considered [3]. APEN Plus was used to simulate the reaction and purification systems of both methods, and alongside literature values and plant cost estimation techniques, the capital and operational expenses were calculated. Finally, the feasibility of both projects is assessed on carbon sequestration mass balances and economic metrics. The main sensitivity variables that need to be improved in order to obtain a minimum ROI threshold value were also evaluated.

Key words: carbon management, process design, microwave plasma reactor

[1] EPA. *Overview of Greenhouse Gases.* United States Environmental Protection Agency. 2017.

[2] Wang, L., Yi, Y., Wu, C., Guo, H., Tu, X. *One-Step Reforming of CO2 and CH4 into High-Value Liquid Chemicals and Fuels at Room Temperature by Plasma-Driven Catalysis.* Angewandte Chemie. 2017.

[3] Lopez, L., Velasco, J., Montes, V., Marinas, A., Cabrera, S., Boutonnet, M., Jaras, S. *Synthesis of Ethanol from Syngas over Rh/MCM-41 Catalyst: Effect of Water on Product Selectivity.* Catalysts. 2015.

589668 Model Predictive Control Design for the Allam Power Plant Integrated with an Air Separation Unit

Abstract:

Since electricity and heat generation account for around 25% of this greenhouse gas emission, high efficiency power generation technology such as the Allam cycle which emits essentially no CO_2 and NOx is much needed. Being an oxy-fuel power cycle, the Allam power plant is inherently integrated with an air separation plant (ASU). In this study Aspen Plus Dynamics was used to model the integrated Allam power plant/Air Separation Unit (ASU) with a high degree of heat and work integration based on the steady state work of Net Power. In our previous work, regulatory and advanced PID controls were implemented and tested for various set point tracking and disturbance rejection scenarios. However, PID controllers have limitations for various reasons. Consequently, a model predictive control (MPC) was designed in this study to optimize and improve the performance of the integrated unit and to test its performance in terms of dynamic behavior and robustness to model uncertainties.

589672 Explore the Function of Ion-Exchange Membrane (IEM) in Membrane Capacitive Deionization (mCDI) Cell, an Alternative Technique in Brackish Water Desalination Field

Abstract:

In the arid west, the freshwater supply of many communities is limited, leading to increased interest in tapping brackish water resources. Although reverse osmosis is the most common technology to upgrade saline waters, there is also interest in developing and improving alternative technologies. Here we focus on Capacitive Deionization (CDI), which has attracted broad attention as an energy efficient desalination technology, especially for ion extraction from low salinity brackish water. The objective of this work is to explore the function of ion-exchange membrane (IEM) in enhancing the performance of CDI cell. In this study, a fully coupled two-dimensional model describing ion transport and adsorption behavior in Membrane Capacitive Deionization (mCDI) cell was developed and solved numerically, taking advantage of Darcy's Law, Nernst-Planck equation, and Donnan equilibrium considering hydraulic dispersion in channel, non-ideal membrane, and modification of ion diffusivity in porous media. The model behavior was supported by the experimentally observed variations in transient effluent concentration under different operating conditions. The results showed that ion flux was significantly improved because of the IEM, leading to a faster desalination rate. The equilibrium adsorption amount is increased due to the permselectivity of the IEM, resulting in extra ion accumulation in electrode's macropores during desalination process. The cut-off time (optimal time to switch polarity to generate highest energy efficiency) of mCDI was shorter with a higher salt removal percentage compared to CDI under the same operating conditions. A sensitivity study of the thickness and the fixed charge density (starting from 500mM) of the IEM was done separately, and neither of them were found to be the dominant influencing factors under a low concentration range of brackish water (below 100mM) in this study.

590103 An Integrated Process for the Production of Propylene Glycol from Shale Gas

Abstract:

Propylene glycol (PG) is produced on a large scale primarily for the production of environmentally safest and least toxic industrial and domestic products such as polyester resins, thermoset plastics, polyurethanes, food additives and antifreeze. In this research, a novel integrated plant is developed for the production of PG from shale gas. This integrated approach has the benefit of safer operating conditions since the intermediate propylene oxide, which is explosive, does not need to be stored and transported. Furthermore, there are potential economic benefits from integration. The overall plant is simulated in the Aspenplus environment, and a variety of process conditions are tested at steady-state to optimize the production of PG. Heat-integration tools are utilized for energy-saving and capital cost reduction opportunities. A comparative economic assessment based on existing plant information indicates that the use of process integration techniques has the potential to reduce costs significantly.

Keywords: Process Integration, Heat Integration, Propylene Glycol

592456 Monitoring Methods for Fault Detection and Diagnosis: Algorithms and Applications

Abstract:

Many industrial processes collect an abundance of data from different sensors. These sensors often measure a wide variety of physical properties in order to ensure that these parameters are adhering to expected values. This is essential to ensure plant and operator safety, increase economic benefits, and maintain product quality.

A number of algorithms have been developed in order to improve existing fault detection and diagnosis performance. These algorithms integrate a number of different data-driven driven tools and methods. Multiscale wavelet-based representation of data can be used in order to handle data that is autocorrelated, non-Gaussian, and noisy. Hypothesis testing methods such as the Generalized Likelihood Ratio (GLR) technique can be used in order to provide the best possible detection for a fixed false alarm rate. Moreover, certain model-based methods have been developed in order to monitor process drifts and degradations in the process model, even when a process is operating under control.

This work will discuss the different tools utilized in development of the algorithms, their features, and practical applications, including the benchmark Tennessee Eastman Process (TEP).

589902 Selective Separation of Industrial Relevant Metal Ions from High Tds Water Using Wafer-Enhanced Electrodeionization (WE-EDI)

Abstract:

There is a need for an efficient and economical process of ion-selective separation. In hydraulic fracturing, for instance, several million gallons of water are used in a site, most of which are disposed of by deep well injection taking the water from the ecosystem and causing earthquakes. However, due to the high amount of ions in wells at fracking sites, pressure driven processes for water recovery are impossible to use. This project will demonstrate an emerging environmentally friendly Wafer Enhanced Electrodeionization (WE-EDI) technology which attempts to eliminate the shortcomings that come with normal EDI and separates effluent streams into concentrated ions and clean water. In WE-EDI, ion-exchange resins are replaced by wafers that consist of cation and anion exchange beads and a polymer binding agent. In this study, selective separation of these ion exchange wafers and the effect of electroosmotic drag on the membrane structure have been performed for various anions and cations (Na+, K+, Ca2+, Cl-) and improved selectivity and higher water recoveries have been observed while significant modeling has taken place to predict the selective separation as a strong function of the equilibrium constants between the ions and the beads. This project is supported by the Membrane Science and Technology (MAST) Center and is conducted with the collaboration and guidance of industry mentors.

593303 Influence of Particles on Amine Losses during CO_2 Capture: A Process Simulation Coupled Aerosol Dynamics Model _

Abstract:

Influence of Particles on Amine Losses during CO_2 Capture: A Process Simulation Coupled Aerosol Dynamics Model

David Dhanraj and Pratim Biswas

Aerosol and Air Quality Research Laboratory

Department of Energy, Environmental and Chemical Engineering

Washington University in St. Louis

St. Louis, MO 63130, USA

Efforts to reduce the concentration of atmospheric CO_2 have been initiated due to escalating concerns of global warming and climate change. Amine-based post-combustion CO_2 capture offers an economically viable retrofit for power-plants in combating these emissions. However, solvent losses have been a major challenge to commercializing this technology. Recent experiments[1] show that vapor-based and aerosol-based solvent losses are the key mechanisms of amine solvent loss during CO_2 capture. While vapor-based amine losses can be abated by employing techniques like water-washing, there has been little work to curtail aerosol-based solvent losses. This is primarily because the underlying mechanisms governing these losses are still not well understood. Although there are a few numerical models proposed for this system, there are no models that consider aerosol phenomena such as coagulation or particle loss. Moreover, these models have not been fully validated through experiments.

Furthermore, there are a few techniques that are being developed to tackle solvent losses. It is known that reducing particle number concentration in the flue gas will reduce amine emissions. Conventional Electrostatic Precipitators (ESP) are however inefficient in capturing particles in the size range 0.1 – 1.0 μm. Amongst different approaches demonstrated, introducing a pre-charging stage and a photo-ionization stage has shown potential in increasing the charging efficiency.

Therefore, the broad objectives of this work are primarily focused on two facets of the amine loss problem. The first is the development of a model that integrates aerosol phenomena with a process engineering model that accounts for the reactive mass transfer process in a packed-bed absorber. The second is the design, fabrication and testing of a bench-scale Photoionization Enhanced Two-staged Electrostatic Precipitator (PI-ESP) at Abbott Power Plant. In essence, with the PI-ESP aimed at reducing the particulate concentration in the flue gas, and the model relating the influence of particulate concentration in the flue gas to amine emissions, this work aims to illuminate the underlying mechanisms governing amine emissions, and demonstrate a solution to tackle the problem.

References

1. International Journal of Greenhouse Gas Control, vol. 85, pp. 86-99, 2019.
2. International Journal of Greenhouse Gas Control, vol. 19, pp. 138-144, 2013.

590287 New TREND: Ethane Transportation with LNG, a Case Study

Abstract:

Demand for ethane for ethylene production is growing worldwide while the USA and other LNG producing regions have surplus ethane available. On the other hand there is a severe short fall of ethane in many countries producing petrochemicals. ***This paper describes a significant market opportunity for transporting ethane co-mingled with LNG and Ethane recovery at the LNG regasification terminal.***

The paper discusses the economics of transporting Ethane with LNG as well as technical challenges in recovery of ethane from LNG. Several different configurations of the regasification terminal and options to minimize changes and cost including Technip Energies Dual Column process (US Patent 7,530,236 B2) are presented.

A comparison is also illustrated with an alternate route of pure ethane liquefaction & transportation which provides marketing flexibility but requires significantly higher investment cost for dedicated terminals and shipping.

590321 Integration of Ejector-Based Flare Gas Recovery and Desalination Processes for Produced Water Onsite Treatment in Shale Gas Production

Abstract:

Flaring is a major concern in the oil & gas industry because it wastes valuable raw materials/energy and emits a substantial amount of air pollutants. Also notable, the produced water treatment during oil and gas production is troublesome due to its large processing volume, high salinity, and expensive disposal cost. Thus, the on-site integration of flaring gas recovery and desalination processes for handling produced water could not only monetize flare emission sources, but also generate freshwater for versatile usages. In this paper, a novel process integrating the ejector-based flare gas recovery process with the thermal vapor compression (TVC) based desalination process (EFGR-TVC) has been developed. It replaces the compressor by an ejector in the flare gas recovery system to power the TVC desalination process more efficiently. The proposed study includes three progressive stages of work: (i) creating a process simulation for the EFGR-TVC process; (ii) operating sensitivity analysis; and (iii) performing an economic evaluation and comparison. In the first stage, rigorous modeling and simulation are conducted for the newly developed EFGR-TVC process. In the second stage, critical design variables affecting the thermal efficiency as well as the system performance are examined, i.e., the sensitivity study for energy intensity (EI) and specific area (SA) for heat transfer with feed water salinity (X), compression ratio (CR), and motive pressure (Pm) will be investigated. In the third stage, fixed and operating costs are examined and compared with the compressor-based TVC process. The simulation results have shown that the replacing the compressor by an ejector in the flare gas recovery system to power the TVC desalination process will be more efficient and economical at normal operating conditions. The EFGR-TVC process could save about 33% capital cost and 16% operating cost per year compared with the compressor-based flare gas recovery TVC process. This study provides a valuable conceptual design to help the oil & gas industry solve both flare emission and produced water treatment issues simultaneously and cost-effectively.

590089 High-Capacity Methane Storage in Flexible Alkane-Linked Porous Aromatic Network Polymers

Abstract:

Adsorbed natural gas (ANG) technology is a viable alternative to conventional liquefied or compressed natural-gas storage. Many different porous materials have been considered for adsorptive, reversible methane storage, but fall short of the US Department of Energy targets (0.5 g g^{-1}, 263 l l^{-1}). Here, we prepare a flexible porous polymer, made from benzene and 1,2-dichloroethane in kilogram batches, that has a high methane working capacity of 0.625 g g^{-1} and 294 l l^{-1} when cycled between 5 and 100 bar pressure. We suggest that the flexibility provides rapid desorption and thermal management, while the hydrophobicity and the nature of the covalently bonded framework allow the material to tolerate harsh conditions. The polymer also shows an adsorbate memory effect, where a less adsorptive gas (N_2) follows the isotherm profile of a high-capacity adsorbate (CO_2), which is attributed to the thermal expansion caused by the adsorption enthalpy. The high CH_4 capacity and memory effect make flexible porous polymers promising candidates for ANG technology.

591624 Syngas Processing Panel Session

Abstract:

The Panelists will discuss Technology Development, Pilot Plant Design and Operation, Technology Commercialization, Operating Plants Upgrading/ Modernization and Feed-Stocks to Chemicals and Petrochemicals.

590372 Drm+Cosorb Process for Syngas Production with High H2/CO Ratio

Abstract:

Dry Reforming of Methane (DRM) refers to the reaction between CH4 and CO2 to produce syngas. However, due to the stoichiometry of the reaction, the H2/CO ratio of the syngas produced is slightly below 1. Life cycle analysis work done by our group [1] indicates that parallel combinations of a DRM unit with a commercial process like Steam Methane Reforming might not result in an appreciable decrease in overall CO_2 emissions or operating costs. Hence, other process combinations were investigated. Out of those processes studied, the DRM+COSORB option gave the best results in terms of reduction of carbon footprint as well as competitive operating costs. The DRM+COSORB process is a stand-alone DRM reformer followed by a CO absorption unit to boost DRM syngas ratio. This process combination has considerable improvement at higher syngas ratios. For a syngas ratio of 2, the DRM+COSORB process would have ~68% reduction in CO_2 emissions and ~21% reduction in operating costs when compared to a Partial Oxidation plant (which utilizes oxygen). Similarly, for a syngas ratio of 3, when compared to an Auto-Thermal Reformer, the reduction in CO_2 emissions is ~72% and reduction in operating costs is ~25%. The current work build on these new targets and by combining information of OPEX and CAPEX of each process and gives more accurate estimates of the economics of each process and their impact on carbon footprint. The proposed DRM+COSORB process will be able to produce high quality syngas, through a stand-alone DRM unit without the need for external hydrogen blending for syngas ratio adjustment. This simulation work of the DRM+COSORB process is one step towards better understanding of this process which has the potential of producing same quality of syngas as commercial processes ($H_2/CO \geq 2$) with lower CO_2 emissions and competitive operating costs.

References:

[1] Afzal et al., *ACS Sustainable Chem. Eng.* 2018, 6, 6, 7532-7544

ID# 589768

Iron supported on natural zeolite (Clinoptilolite) as a low-temperature Fischer-Tropsch synthesis catalyst

Roick Chikati[1], Diakanua Nkazi [1], Joshua Gorimbo[2]*

[1]Department of Chemical and Metallurgical Engineering, University of the Witwatersrand, Private Bag 3, Wits 2050, South Africa
[2]Institute for the Development of Energy for African Sustainability (IDEAS), University of South Africa (UNISA), Florida Campus, Private Bag X6, Johannesburg 1710, South Africa

**Corresponding Author Email address: joshuagorimbo@gmail.com*

Abstract

A fundamental difference exists between commercial supports such as silica and alumina with functional porous materials and natural supports such as zeolites. In this study natural zeolite called clinoptilolite (a type of zeolite found in South Africa) was used as a catalyst support. This support proved to be effective for low-temperature and pressure Fischer–Tropsch synthesis (LTFTS). Synthesis of this catalyst was by loading of iron on clinoptilolite through the wet impregnation method. The catalyst was then loaded into the reactor and reduced with hydrogen prior to FTS. The effects of its use as support in FTS were investigated in a fixed bed reactor. It was found that the CO consumption rate of 1.02 x 10^{-4} mol/min.gcat of which 7.24 x 10^{-5} mol/min.gcat was the actual Fischer Tropsch rate with the remaining 2.93 x 10^{-5} mol/min.gcat consumed by the WGS reaction. The product distribution of the gaseous phase analysed were more olefinic than paraffinic. The product distribution for our experimental condition agreed with a one alpha ASF distribution with an alpha value of 0.86. These findings may permit the development of new effective support materials, which are cost effective for clean fuel production via FTS process.

Keywords: Clinoptilolite, zeolites, iron catalyst, Fischer Tropsch synthesis, product distribution, porous materials

1. Introduction

Research conducted thus far has endeavored to optimize each FTS stage in an attempt to improve production and lower the operation cost. Furthermore, the literature shows that several studies have focused on the effect of FT operating parameters, including feed sources, type catalysts and catalyst supports [1–7]. This study intends to extend the investigation and focus specifically on the effect of catalyst supports in a naturally available material (zeolite) for FT process. The choice of the catalyst is governed by factors such as cost and the selectivity towards preferred products. Iron and cobalt are the most used metal-based FT catalysts in the industry. For instance, currently all the commercial Sasol FT operations are using Fe-based catalysts, which are synthesized onsite [8]. Iron based catalysts have been commonly used as they are comparatively cheaper and readily available.

Several inorganic oxides eg. silica (SiO_2) [9] , alumina (Al_2O_3) [10], titania (TiO_2) [11] and silicon carbide [12] have been thoroughly researched as supports for Fe catalysts. Nonetheless, significant support-metal interactions during reduction have been observed the trend: $Al_2O_3 > TiO_2 > SiO_2$ for cobalt metal[13]. These interactions often make the catalyst difficult to reduce. Addition of reduction promoter is therefore done to enhance catalyst reducibility. X- ray fluorescence studies done on South African zeolite reveals that the zeolite consists of all the phases of material used as supports and promoters in FTS catalyst [14]. Adding supporters to aid catalyst reducibility and activity metals such as alkali metals would not be necessary since they are already contained in the natural zeolite. South African zeolite contains promoter percentages that are almost identical to those recommended in FTS catalyst [15].

Catalyst support is as important as the metals deposited on them, as this influences several factors including the activity, selectivity, catalyst stability and the costing of the catalyst. However, presently most of the catalyst supports used in FTS are commercially synthesized. Natural zeolites (Clinoptilolite, which is mined in South Africa) have the physico-chemical properties of a good support, an adequate phase makeup and other constituents congruent with commercial supports. It is therefore warranted that a comprehensive investigation is undertaken to shed some light on natural zeolite supported FT catalysts. As a result, the cost of energy from FTS could perhaps be lowered below the conventional generating systems despite the benefits of higher efficiency and cleaner environment.

2. Experimental

Catalyst preparation

The Clinoptilolite supported iron (Fe/Clino) catalyst was prepared by depositing an iron precursor component onto the clinoptilolite support from the precursor solution using incipient wetness impregnation method as reported in the literature [16,17]. The zeolite (Clinoptilolite) support was first pre-treated by mixing it with distilled water in a mass ratio of 1:1, and air dried at 120 ^{0}C for 1 hour. This was then followed by 16 hours calcinations at 400 ^{0}C. The calcined material was then crushed and sieved resulting in particles with diameters between 0.5–1 mm. Thereafter, the treated support was impregnated with Fe $(NO_3)_3{\bullet}9H_2O$ solution sufficient enough to give an iron metal loading of ca. 14% by mass. After impregnation, the impregnated material was then dried in air at 120 ^{0}C for 16 hours and again calcined in air at 400 ^{0}C for 6 hours. This process allowed the iron nitrate to decompose and transform to iron oxide, as well as removing solvent and/or water therefrom - resulting in the formation of the Fe-Clino catalyst. The most attractive feature of this method of preparation is its simplicity of practical application. The method is also comparatively economical and has been employed by several Fischer Tropsch practitioners [18–20].

FT synthesis

Fischer-Tropsch reaction started after the reduction when the pressure of the reactor was increased from atmospheric to 1.85 bar(abs) while the feed (syngas) continued to flow in the system. The FTS experiments were carried out under the reaction conditions similar to those reported by Gorimbo et al, [20] for **450** hours of time on stream (TOS). The results obtained from these FTS runs are presented and discussed. The syngas feed composition was 60% H_2, 30% CO and 10% N_2.

3. Results and discussion

Catalytic activity

The catalyst activities measured by CO conversion, water gas shift (WGS) selectivity and FT selectivity as function of TOS during FTS is shown in **Figure 1**. Separatly, the water-gas shift (WGS) selectivity is the ratio of CO_2 production rate to total CO rate and FT selectivity was determined as a function of CO converted to hydrocarbons. The three graphed parameters showed desirable distribution with the WGS and FT selectivity of 19.02±2.61 and 80.98±2.61 respectively for conversions averaging 11.09±0.49%.

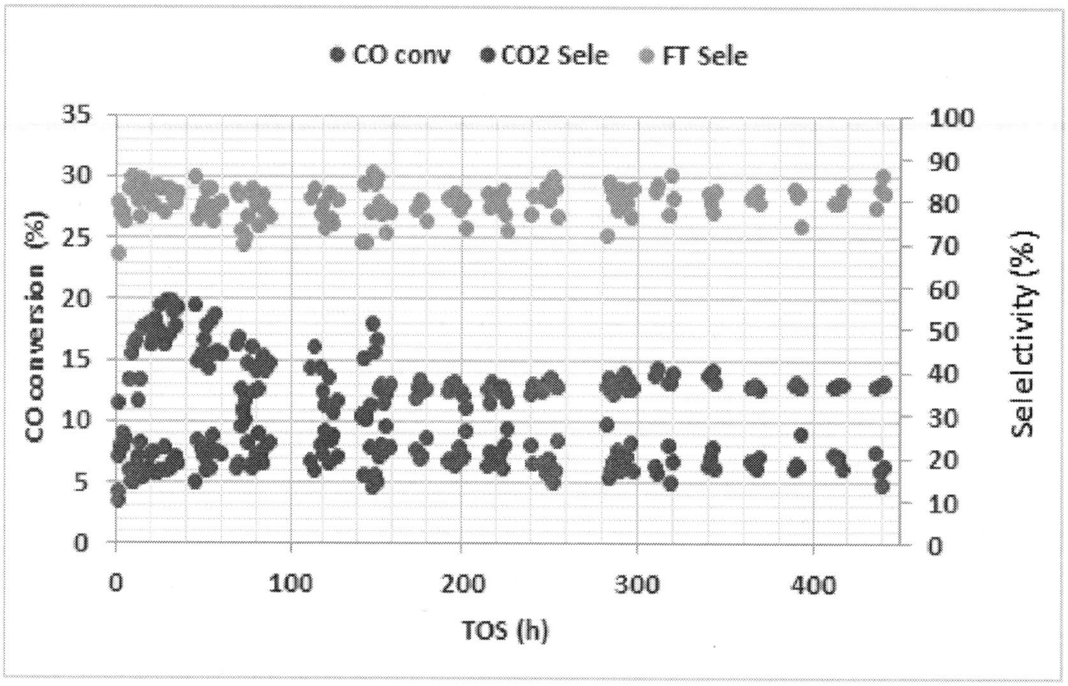

Figure 1: CO conversion, WGS selectivity and FT selectivity under operating conditions: 250 °C, 2592 h^{-1} at 1.85 bar (abs).

An increase in WGS selectivity can be used as a measure of catalyst deactivation by oxidation as magnetite is known to be the active catalyst for WGS reaction[21–23]. In this study, the deactivation of the catalyst was not observed. Instead the results showed that the catalyst activity in particular the Fischer Tropsch selectivity was comparatively higher than water gas shift reactions for Fe-Clino catalyst.

Rates

Water is produced as a by-product in FT reaction, which is further consumed in the WGS reaction, thus the WGS reaction rate is controlled by the amount of water formed in the FTS. Since water is not furnished to the reaction, the stoichiometry of the FTS and WGS reactions agrees with the following condition:

$r_{WGS} \ll r_{FTS}$ Equation 1

The production of the undesired CO_2 is as a result of the presence of magnetite fractions in the catalyst bed which catalyse the Boudouard reaction. The behaviour of Fe-Clino catalyst used in these experiments behaved the same from the other used iron catalyst based on different supports for instance, the 10% Fe loading on Silica reduced with H_2 reported in studies by Gorimbo et al, gave almost the same conversion [24].

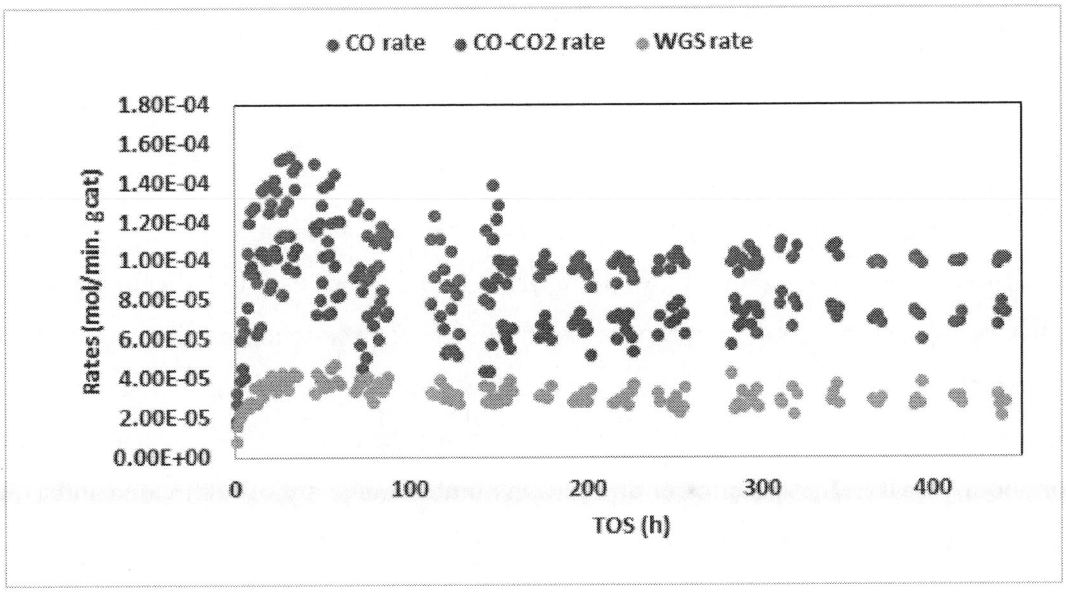

Figure 2: CO consumption rate, WGS rate and FTS rate under operating conditions: 250 °C, 2592 h-1 at 1.85 bar (abs).

From **figure 2**, the averaged CO consumption rate, WGS rate and FT rate with standard deviation errors obtained after 450 h TOS as $1.02 \pm 0.037 \times 10^{-4}$, $1.02 \pm 0.037 \times 10^{-4}$, and $1.02 \pm 0.037 \times 10^{-4}$ mol/min.gcat respectively.

From the consumption rates of CO, the paraffin production rate and olefins production rate are tabulated below (Table 1) for the TOS of 0 to 450 h.

Table 1: Averaged paraffins and olefin production rate with standard deviation errors obtained from 0 to 450 h TOS

Species	Production rate±SD (mol/min.gcat)	
	Paraffins	Olefins
C2	$(3.93 \pm 0.77) \times 10^{-7}$	$(9.83 \pm 1.91) \times 10^{-7}$
C3	$(1.40 \pm 0.15) \times 10^{-7}$	$(1.76 \pm 0.09) \times 10^{-6}$
C4	$(4.67 \pm 0.35) \times 10^{-7}$	$(9.82 \pm 0.40) \times 10^{-7}$
C5	$(1.62 \pm 0.11) \times 10^{-7}$	$(7.14 \pm 0.33) \times 10^{-7}$

The trend that can be deduced from the values in Table 1 is as follows: $C_3 < C_2 < C_4 < C_5$ paraffin trend and $C_5 < C_4 < C_2 < C_3$ Olefin trend. These trends clearly indicate the negating behavior in the production of paraffins and olefins. The production rates of paraffins tend to increase from C_2 to C_5 and an opposite trend is observed for olefins. The recorded CH_4 content was fairly low, while the light hydrocarbon content in the C_2–C_4 range accounted for about 15 wt%. The α-olefins in this carbon number range are of high value and usually used as feedstock in industries.

4. Conclusions

The primary objective of the research undertaken for this work was to utilize the experimental findings to advance knowledge on the possibilities of using naturally occuring zeolites as catalyst support in FTS. Substantial progress has been made in this work towards generating detailed knowledge on the use of clinoptilolite as a support.

Experimental investigations were performed at laboratory scale using a fixed bed reactor and pressure 1.85 bar (absolute) and gas hourly space velocity of 2592 h^{-1} and fixed reactor temperature of 250 °C, where invaluable data were successfully collected and analysed for a TOS ca.450 h. The experimental work proved the possibility of the use of natural zeolites as supports. The data collected in this project, though quite useful, remains valid only to the fixed bed type of reactors investigated. However, it is hoped that the project has given some impetus for further research, and where the analysis is inconclusive, it provokes further thinking for improvements. Thus, this work is a contribution to the understanding of the FT process at using clinoptilolite as a support for iron-based catalyst.

Acknowledgment

The authors are grateful for the financial support provided by the University of the Witwatersrand, University of South Africa (UNISA), National Research Foundation (NRF) of South Africa, and the Institute for the Development of Energy for African Sustainability (IDEAS) research unit at UNISA.

References

[1] B. Bao, M.M. El-Halwagi, N.O. Elbashir, Fuel Process. Technol. 91 (2010) 703–713.

[2] F. Pardo-tarifa, Cobalt Catalyst Supports for Fischer-Tropsch Synthesis, 2017.

[3] D.A. Wood, C. Nwaoha, B.F. Towler, J. Nat. Gas Sci. Eng. 9 (2012) 196–208.

[4] J. Gorimbo, Catal. Sci. Technol. 8 (2018) 2022–2029.

[5] F.E.M. Farias, F.A.N. Fernandes, F.G. Sales, Lat. Am. Appl. Res. 40 (2010) 161–166.

[6] H. Atashi, S. Razmjooei, M. Khorashadizadeh, M. Shiva, F.F. Tabrizi, S.A.H.S. Mousavi, J. Taiwan Inst. Chem. Eng. 54 (2015) 83–90.

[7] H. Atashi, S. Razmjooei, M. Khorashadizadeh, M. Shiva, F.F. Tabrizi, S.A.H.S. Mousavi, J. Taiwan Inst. Chem. Eng. 54 (2015) 83–90.

[8] L.P. Dancuart, A.P. Steynberg, Stud. Surf. Sci. Catal. 163 (2007) 379–399.

[9] P.A. Chernavskii, V.O. Kazak, G.V. Pankina, V.V. Ordomsky, A.Y. Khodakov, ChemCatChem 8 (2016) 390–395.

[10] J. Zhang, J. Chen, J. Ren, Y. Li, Y. Sun, Fuel 82 (2003) 581–586.

[11] S. Bagheri, N.M. Julkapli, S. Bee, A. Hamid, Sci. World J. 2014 (2014).

[12] M.J. Ledoux, C. Pham-huu, Featur. Lab. Des Matériaux, Surfaces Procédés Pour La Catal. ECPM 5 (2001).

[13] G. Jacobs, T.K. Das, Y. Zhang, J. Li, G. Racoillet, B.H. Davis, 233 (2002) 263–281.

[14] J. Gorimbo, B. Taenzana, K. Kapanji, L.L. Jewell, South Africa J. Sci. 110 (2014) 1–7.

[15] J. Gorimbo, Effect of the Homoionic Form of Clinoptilolite on Ni2+ Adosorption Isotherms: A Thermodynamic Study, 2011.

[16] H. Li, B. Hou, J. Wang, C. Qin, M. Zhong, X. Huang, L. Jia, D. Li, Mol. Catal. 459 (2018) 106–112.

[17] K. Keyvanloo, B. Huang, T. Okeson, H. Hamdeh, W. Hecker, Catalysts 8 (2018) 77.

[18] A.A. Muleja, Study of the Selectivity To Light Hydrocarbons in Fischer-Tropsch Synthesis, University of the Witwatersrand, 2016.

[19] J. Gorimbo, An experimental and thermodynamic study of Iron catalyst activation and deactivation during Fischer Tropsch synthesis, University of the Witwatersrand, 2016.

[20] J. Gorimbo, X. Lu, X. Liu, D. Hildebrandt, D. Glasser, Appl. Catal. A Gen. 534 (2017) 1–11.

[21] E. de Smit, B.M. Weckhuysen, Chem. Soc. Rev. 37 (2008) 2758–2781.

[22] K. Keyvanloo, W.C. Hecker, B.F. Woodfield, C.H. Bartholomew, J. Catal. 319 (2014) 220–231.

[23] D.B. Bukur, L. Nowicki, S.A. Patel, Can. J. Chem. Eng. 74 (1996) 399–404.

[24] J. Gorimbo, A. Muleja, X. Liu, D. Hildebrandt, Int. J. Ind. Chem. 9 (2018) 317–333.

AUTHOR INDEX